Urban Norlén

Simulation Model Building

Simulation Model Building

A Statistical Approach
to Modelling in the Social Sciences
with the Simulation Method

by

Urban Norlén

A Halsted Press Book
John Wiley & Sons
New York–London–Sydney–Toronto

Library of Congress Catalog Card No. 75–4935

ISBN 0-470-65090-7

Printed in Sweden by

Almqvist & Wiksell, Uppsala 1975

Preface

The present work deals chiefly with theories for, and applications of, models based upon the use of simulation in the social sciences. In this connection interest has been focussed mainly upon the statistical aspects of the subject.

Simulation as a research tool seems nowadays to be attracting the attention of social researchers more and more, and its utilization is becoming ever more widespread.

This development has to a large extent been promoted by the technical build-up of high-capacity computers and by the emergence of specially adapted simulation programming languages. The work with simulation models has also been able to reap the gains due to the appearance of interdisciplinary sciences such as cybernetics and the systems theory.

For a description of dynamic processes in real systems the properties pertaining to the simulation concept appear to be well suited. These possibilities should be regarded as given improvements and they extend the scope of statistical procedures implying advantages for the model builder. But the change—it should be pointed out—is a double-edged matter, as increasing difficulties in understanding go hand in hand with the increasing complexity of the model.

The above consideration led to the conclusion that a relatively urgent interdisciplinary research contribution should be initiated with the aim of providing a statistical frame of reference for work with simulation models so as to make it possible for us to implement the information that can be conveyed by them.

Of the persons with whom I have had to do in connection with my research based on the above view of the problem and presented in this work, I wish to mention the following.

Professor Herman Wold, who has been my teacher and guide, has encouraged me to make a study of the methodology of simulation in the subject statistics. As the work has proceeded he has with unswerving interest followed the development of the project and has throughout given me invaluable support and guidance, not least in those questions having to do with the interdisciplinary alignment of the task.

Docent Stein Bråten, who is my co-author in the Appendix to this dissertation, has introduced to me problems associated with simulation models. In stimulating discussions of problems of method arising in this connection he has given me discerning and witty assistance.

In clarifying discussions Docent Ejnar Lyttkens has with interest and personal engagement deepened my insight in matters pertaining to statistical problems. Fruitful discussions of statistical problems have also been carried on with FK Jan Selén.

Professor Tore Dalenius has kindly given his scrutinizing attention to the views concerning topical survey problems advanced in Chapter 7.

On Chapter 9, dealing with pedagogical applications, Docents Gösta Berglund and Ulf Lundgren have made valuable comments. At the symposium "Evaluation of Education" held in Uppsala on November 1st–2nd 1971 Professors Karl-Georg Ahlström and Urban Dahllöf gave me the opportunity to present the main points in this chapter.

To these persons I wish to convey my deep gratitude. I am also indebted to those of my friends and colleagues who at seminars and during other discussions have assisted me in my work with advice and points of view that have influenced me in the framing of the work in a positive direction.

My thanks for able help in the translation of my thesis into English go to Mr Donald Burton, David Edgerton, B. Sc., and FM Lennart Littorin.

For their typing of the manuscript of the thesis with such understanding and patience I must express my gratitude to Mrs Ingrid Björklund and FK Imbi Seiman.

I have had, from first to last, the strong support of my father, Mr Arthur Norlén, who helped me to organize the work and suggested stylistic improvements in the Swedish manuscript. — The book is dedicated to him, and to my wife Agneta and my mother, to both of whom I am deeply indebted for their support. These three persons have made this study possible.

Needless to say, the responsibility for any mistakes and errors in the study is solely mine. Furthermore, owing to the interdisciplinary character of the contribution, the study may contain ideas which are not suitable for some real systems. The best way to obtain a more solid foundation and more developed principles is to get a more comprehensive material of applied work. In this respect the present work should be regarded rather as a status report.

U. N.

Contents

PART II: APPLICATIONS

Chapter 8. Alignment of the applications presented

Chapter 9. Simulation modelling of education

Appendix. Description and models of a simulation of communication model (with Stein Bråten)

ERRATA

P. 45: 3–4 "determinate system with combinatorial behaviour" read "system with random behaviour"
P. 46: 9 below fig. $g(\eta|\xi)$, read $g(\eta|\zeta)$
P. 46: 10 below fig. expectations, read distributions
P. 59: 7 from below $E(\varepsilon)'E(\varepsilon)$, read $E(\varepsilon)'D(\varepsilon)^{-1}D(\varepsilon)$
P. 65: 9 Students', read Student's
P. 68: 6 from below of the order N consisting, read consisting
P. 71: 20 (6.2.13), read (6.2.12)
P. 73: 20 $=Aa$, read $=Aa^{\Delta}$
P. 78: 13 full, read "multiple" of a full
P. 79: 4 (relation (6.5.26)) $x_i=\{$, read $x_i=x_i^0\{$
P. 95: 6 from below (7.2.34), read (7.2.33)
P. 97: 7 from below eigen-value one, read eigenvalue-one
P. 97: 4 from below eigen-values, read eigenvalues
P. 144: 3 Könsegrering, read Könssegregering
P. 145: 9 from below Metodologie, read Methodologie
P. 145: 6 from below Humbolt, read Humblot

Chapter 1

Presentation of the research project

1.1. Problem situation and aims of the project

The point of departure for this research project is constituted by methodological problem-complexes of a chiefly statistical nature which arise in connection with the computer use of simulation techniques in model building in the social sciences.

In the account of the research problem use has been made of Bunge's classification of scientific problems as substantial or object problems on the one hand and strategy or problems of procedure on the other hand (Bunge 1967: I).

To begin with, the object problems and the method for solving them considered in this study may be focussed with the help of the following quotation: "... if the investigator's problem area happens to involve what might be called a *middle-level* problem, that is, a problem too complex to be handled by traditional methods but not so global as to defy analysis, then computer simulation may be appropriate" (Abelson 1968, p. 283).

Hence, the assumption in cybernetics of the complexity of biological and social systems takes a prominent place among the background factors in this study; cf. Ashby (1956). The distinction emerging from the designation "middle-level problem" makes it clear that whereas the original system is complex, it is assumed that this system can be structured in meaningful categories that make it possible to analyse separately each of the categories set up as well as to analyse them in relation to one another. It is here that simulation enters and indicates a way of collocating syntheses, i.e. taking the results back to the original problem level.

The above description indicates, in other words, that simulation can serve as an aid in the bridging over of differences between existing object problems that cannot be resolved with traditional methods.

The following quotation exemplifies a problem of procedure that is of topical interest in connection with the use of simulation: "... the problem of validating computer models remains today perhaps the most elusive of all the unresolved methodological problems associated with computer simulation techniques" (Naylor 1971, p. 153). Apart from the mention of the problems connected with the matter of validation—which have for the most part to do with the direct confrontation of computer models with empirical facts—no broad account of statistical work with the types of model in question has as far as I am aware been reported or for that matter been completely developed.

There are thus grounds for assuming that if the problems of procedure connected with the use of simulation can be solved, this should lead to a wider field of application for models. It is against this problem background that the aims of the research project have been given the following alignment.

The target set up for the project is to develop a statistical method for the modelling of complex social systems. In the project, attention is focussed upon the application of statistical principles to the development of formalized models in the social sciences based upon the computer use of the simulation technique for the purpose of synthesizing theoretical/empirical knowledge. Especially is attention concentrated upon the questions connected with the testing and validation problems.

1.2. Direction of research

Computer models do not in principle differ from other formalized models. It should therefore be possible to develop the work on the basis of the accumulated knowledge in appropriate fields. Of the attempts made in connection with such an alignment attention is drawn first and foremost to the following works.

A. In the general orientation towards the concept of a model and the epistemological questions associated therewith, particular attention is given to the works by Bunge (1967) and Wold (1967*a–b*, 1969*a*).

B. In advancing principles for computer models attention is paid to econometric multi-relation model building, whose foundations are generally ascribed to Tinbergen (1937) and Haavelmo (1943), two pioneers in the development of the science. In the realm of non-experimental model building, econometrics has played a pioneering role and provides a proved frame of reference; cf. Wold (1969*c*).

C. For the specific framing of the concept of a model the point of departure has been constituted by the different kinds of concept in cybernetics referring to the relations between two systems: the formalized model system and the modelled system. The most important reference in this connection is Klir & Valach (1967).

D. In the present work, reference has been made to cybernetics also with respect to some of the trains of thought advanced in this science, including basic concepts, problems and methodological principles as these are presented in two works by Ashby (1952, 1956).

E. Of the material prerequisites in the form of digital computers with programming languages developed for model formulation and for simulation, the SIMULA-language designed by Dahl and Nygaard (1966) is taken as point of departure.[1]

F. In the orientation concerning the mode to be chosen for a confrontation of the model with empirical findings the principles developed in statistical theory and application for the testing of hypotheses, constitute the basis. Already at this stage it should be pointed out that this orientation towards the testing of hypotheses implies consequences not only in test situations, but also in other stages of the modelling work, including the problems of formulation and estimation that are intimately bound up with the initial construction and with the further adaptation of the models.[2]

1.3. Planning of the report

The material in this study is divided into a theoretical part, consisting of Chapters 2–7 and a section devoted to applications, comprising Chapters 8–9 and an Appendix.

The next chapter (2) deals with the foundations for the present modelling approach. First the notions and meanings given to model building and simulation in current usage are introduced. Then a meaning for the model concept is presented with reference to other works. Basic distinctions made are illustrated in the formalization of an exemplified verbal theoretical statement. In a subsequent section a definite purpose is given to the use of the simulation method in model building. Finally, properties of simulation models and the target aimed at are briefly described in relation to fundamental aims in the modelling of social systems.

In Chapter 3 an inventory of the different problem-fields in connection with modelling is made with the systems concept as an organizing device. The following problems are distinguished:

- construction of model systems
- testing of model systems
- analysis of model systems
- analysis of modelled systems
- application of model systems.

Brief accounts are also given of the first and last problems in the above list.

---

[1] Analog or hybrids-analog-digital computation is not considered, since there seems to be little need for it, either as regards aspects having to do with model representation or from computational points of view. It would have been of interest, on the other hand, to include in the study the combination of a computer with oscilloscope displays and other facilities linked to the computer through its input–output. Since this study is mainly focussed on methodological problems, however, such technical aspects have been considered to be outside the scope of this work.

[2] The basic principles of hypothesis testing are generally ascribed to Neyman & Pearson (1933). For a review, see Cramér (1945).

Chapter 4 deals with conceptual aspects and presents the modelling approach. First the structure for the anticipated model systems is described and related to other parallel works. Then the paradigm for simulation with the model systems is described. A subsequent section gives an account of the way in which modelled systems are viewed and interpreted. Finally, the relation between the model and modelled systems aimed at is given in the form of a model specification.

In Chapters 5–7 one chapter each is devoted to the three intermediate problems in the above list: The testing problem, the analysis of model systems and the analysis of modelled systems. The chief emphasis is on the solution of problems which are of particular importance for the types of model considered and which have not previously been dealt with to any great extent.

In Chapter 5 the testing problem is ventilated. Taking as point of departure a model specification, in the main two types of attempts at a solution are distinguished: an implicit approach and an explicit approach. The implicit approach is based on the systematic treatment of general and intuitive judgements concerning the model in the form of so-called Turing tests, while the explicit attempt is based on the so-called goodness-of-fit tests applied to characteristics of model outcome.

In the discussion of the testing problem we are confronted with defined needs for analytic results from both the model and the modelled systems. In addition to this, there is the need for analyses relating to the two systems of a more general and explorative kind.

Chapter 6 is devoted to the question of the analysis of model systems, in which connection the use of the simulation method occupies a central position. At the same time the chapter contains an account of the design and analysis of experiments on ways of ascertaining model properties.

Chapter 7 gives an account of the analysis of modelled systems, the exposition being in the first place devoted to the presentation of an exploratory technique within an econometric framework. This technique is adapted to the problem of finding econometric relations between observed variables showing good agreements in the sense that the residuals become small. In the second place the account is largely expository and directed towards the task of bringing together previously discussed needs for analysis with the theory of parameter estimation and the survey theory.

In Chapter 8, the first in the part devoted to applications, a summarizing account is given of the research direction in the two applications presented in Chapter 9 and in the Appendix.

In Chapter 9 an attempt at modelling a system of education is presented. The presentation gives a more detailed account of the construction work, as this problem-complex is treated rather summarily in the theoretical part of the study. In this connection illustrations are given of the way in which partial theories from different social sciences can be and are used to give substance to the model. Some preliminary results from simulations with the model are reported.

In the Appendix an earlier constructed model of communication in target populations is analysed. In this study the complexity of the model system has proved to require a considerable amount of work to attain an understanding of and the ability to use the types of model discussed. In this connection the question arises as to whether it is possible to substitute for the primary model a simpler model system retaining the most important properties of the original. In the Appendix the question of reduction is discussed with reference to attempts to simplify the model analysed with the aim of producing simpler relation models.

1.4. Principles of notation

The text is accompanied with formulas, definitions, figures, tables, footnotes and examples. The formulas are numbered by chapter and section. Definitions, figures, tables and examples are numbered by chapter.

The symbols used are generally explained in their context.

As a rule Greek letters and Latin letters with some distinguishing marks are used to designate hypothetical parameters and variables. Latin letters are used to designate observed quantities. To the extent deemed suitable, estimates of a parameter or an observation of the value of a variable are designated with the Latin letter corresponding to the Greek letter for the parameter or variable. In clear cases, for stochastic variables, sometimes the same notations will be used for the stochastic variables and values on the variables.

Matrix algebra is frequently used. In general, matrices are designated with medium-faced capitals and vectors with medium-faced small letters. All vectors are column vectors. To indicate row vectors and transposed matrices, accents are used.

If $\mathbf{A}$ is a matrix, a_{rs} is used to designate an element in row r and column s, and $\mathbf{a}_s$ to designate column s in $\mathbf{A}$. The element in row r in the vector $\mathbf{b}$ is indicated with b_r.

If $\boldsymbol{\xi}$ is a stochastic variable, then $E(\boldsymbol{\xi})$ and $D(\boldsymbol{\xi})$ are used to designate the mean-value vector and the dispersion matrix, i.e. the variance-covariance matrix, for $\boldsymbol{\xi}$.

In the case of differential calculus the modes of designation may be exemplified as follows:

$\frac{\partial y}{\partial \mathbf{x}}$ is the vector with element $\frac{\partial y}{\partial x_r}$ in row r

$\dfrac{\partial^2 y}{\partial \mathbf{x}^2}$ is the matrix with element $\dfrac{\partial^2 y}{\partial x_r \partial x_s}$ in row r and column s

$\dfrac{\partial \mathbf{y}}{\partial \mathbf{x}}$ is the matrix with element $\dfrac{\partial y_r}{\partial x_s}$ in row r and column s

$\dfrac{\partial y}{\partial \mathbf{X}}$ is the matrix with element $\dfrac{\partial y}{\partial x_{rs}}$ in row r and column s.

Part I: Theory

Chapter 2

Model building and simulation

2.1. Introduction

In this chapter are discussed the concepts model and simulation. The presentation aims at giving an account of the place taken by simulation in the building of a model. Most aspects emerging in the formation of concepts are best understood in connection with the concrete development of the modelling approach. These aspects are dealt with in later chapters. Here we shall be discussing chiefly the more general properties of the concepts that follow from the alignment of this study in the direction of formalized models given in the form of computer programs.

2.2. The notions of modelling and simulation

The concept of a model and the therewith closely connected concept of simulation have long been familiar in research work. Research may be described as model building (Wold 1967*a*), and the design of a conceptual model has been referred to as modelling (Bunge 1967: II). We shall use these two concepts model building and modelling synonymously in what follows.

In the present work the concept model will conform with Wold's usage. In this connection the concept (cognitive) model is used in the sense of a joint theoretico-empirical construction. The theoretical part is in the nature of things hypothetical, whereas the empirical part consists in the matching with empirical evidence within the frame for hypothesis testing (Wold 1969*a*).

The simulation concept finds its proper place in the experimental work with the constructed model. In this study, as has already been mentioned, modelling is dealt with in the form of mathematically formulated models. In short, simulation is here described as a numerical technique for the carrying out of experiments with the model (Naylor 1971). Other and similar definitions are reported by, inter alios, Forrester (1961) and Newell & Simon (1963).

The content of the concept simulation is in the present work restricted chiefly to the covering of only numerical operations in a model that are performed in a computer with the intention of arriving at model implication. To indicate the difference between simulation experiments with models and experiments carried out in real systems, Bunge has launched the term thought-experiment in connection with his discussion of simulation (Bunge 1967: II). This distinction is of course important, since the two kinds of activity stem partly from different posings of the problem. The term thought-experiment is, however, misplaced inasmuch it can give rise to misinterpretations. So, for example, it does not follow from what has been said that simulation experiments are necessarily a less systematic and more loosely founded procedure for the production of consequences. And whereas Bunge argues for the use of simulation only in applied science, we shall maintain that simulation has an equally important function in pure science.

2.3. The structure and the content of a formalized model

In this section we shall elaborate further what has earlier been said with two new concepts: the structure and content of a model.

In the first place the (logical) structure for a model system constitutes its logical system of symbols regarded without reference to its theoretical and empirical context. In other words, the structure forms a deductive system in itself, in Bunge's terminology an abstract theory (Bunge 1967: I).

And in the second place the content of a model system consists of an interpretation of the symbols used in the structure. These interpretations are of two kinds: referential and evidential (idem).

The referential interpretations are also of two kinds: (1) rules in which the structure's symbols are made to correspond to the (non-formal) theoretical concepts and (2) assumptions in which the symbols are made to correspond to hypothetical (idealized) entities in the modelled system (idem).

The evidential interpretations are assumptions in which the structure's symbols are linked to observable entities in the modelled system. In Bunge's terminology the empirical content or specification of the model used here might be called the operational interpretation of a structure (idem).

The evidential interpretations thus delimit the empirical part of the model, i.e. the model's relation to the modelled system. By the theoretical part of the model we shall in what follows be referring to the theory used and the referential interpretations in the model's relation to theory, i.e. how the theory is actually interpreted; cf. Wold (1967*a–b*, 1969*a*).

The orientation in this work towards formalized models within the frame for the concept of the model advanced calls for some further comments, and this chiefly for the reason that social scientific theories are for the most part verbally formulated. The use of a verbal theory in a formalized model thus seems obviously in the first place to call for a set of interpreted mathematical symbols for the verbal concepts in theoretical propositions.[1]

In the following example the formalization process from the point of departure of a theoretical statement is illustrated.

Example 2.1. Consider a verbal theory about a verbally formulated variable, which is of course not treated in isolation, but together with other, likewise theoretical and verbal, variables. As an example we consider the variables given in italics in: "Persons with high *intelligence* have higher ability to *adapt* themselves purposefully to new *situations*." The variables *intelligence* and *situation* are considered as given or predetermined (explanatory), whereas the behaviour variable *adapt* is considered as the endogenous variable to be explained.

A framing of a model using the distinctions made and based on the above theoretical statement can be exemplified as follows.

The structure of the model. The model structure is specified as the conditional probability distribution $f(\eta \mid \zeta, \xi)$ for a stochastic variable, given the values of the variables ζ and ξ.

The content of the model. The content of the model involves the following three kinds of interpretation:

- The three variables in the model structure correspond to the theoretical concepts: *adapt* (η), *intelligence* (ζ) and *situation* (ξ) (the referential interpretation rules).
- It is assumed that a value for ζ (*intelligence*) can be attributed to a person (the modelled system). When this person is confronted with a situation characterized by a specified value for ξ (*situation*), an observation on the random variable (*adapt*), distributed as $f(\eta \mid \zeta, \xi)$, is obtained, the observation being regarded as the outcome of some (fictitious) experiment (the referential interpretation assumptions).
- Finally, it is assumed that observed values from the modelled system can be obtained as follows. A measure for ζ (*intelligence*) may be defined as the person's response to certain intelligence tests. An observation on η (*adapt*) may then be defined as the person's response according to some classification scheme from an experiment with the value for ξ (*situation*) kept at some specified level according to some quantification rule (the evidential interpretation assumptions).

---

[1] Blalock (1969) has considered the problem of recasting sociological verbal theories to econometric relation structures. His study points out many of the problems inherent in the formalization process, such as e.g. the need for: clarification of concepts, integration of propositions from several sources and the search for implicit assumptions.

The characteristic features that the above illustration is intended to throw light upon imply, to specify the consequences, in the first place that different kinds of concept, both theoretical and empirical, are indispensable in connection with the kind of modelling here referred to. And in the second place that much of the lack of clarity from which model building suffers will disappear if the different forms of concept referred to above are allowed to exist parallel with one another. This view of the problem is in agreement with the discussion carried on by Wold (1969*a*) and others.

The same reasoning and distinctions between theoretical, model and empirical constructions will be used for relations between concepts and the arranging of these relations in systems. The extension of this view appears especially important in connection with the approach to, inter alia, the concept of causality, where the theoretical notion of a cause–effect relation is kept separate from the corresponding given logical structure of the model and empirical concepts; cf. (Wold 1959*a*).

One consequence of what has been adduced above is that one objective in the building of a model will be the setting up and establishment of mutual connections between currently important theoretical, empirical and model systems. As regards the question of the relations between model and modelled systems, the empirical part of the models, we shall be reverting to these in later sections. The matter of mutual connections between theoretical and model systems, which constitute the theoretical aspects, will be given more summary treatment.

It will thus be assumed that the theoretical aspects, although generally implicit, will exist also in subsequent chapters. As has already been mentioned, however, the discussion will for the most part be carried on only in terms which on the one hand include the concepts in the formalized model, and on the other hand the more or less corresponding empirical concepts.[1]

2.4. Simulation as an aid in making syntheses

The use of simulation in making syntheses is elucidated in this section on the basis of the distinction between the structure and content of a model drawn in the foregoing. In accordance with what was mentioned in section 2.2, the concept simulation may include the activities and deductions that are put into the structure of a model with the aim of obtaining new knowledge or

[1] Our eclectic approach emphasizes points of agreement with some aspects of Bunge's and Wold's presentations. While Bunge's model is the application of a theory in the form of a logical structure with ensuing interpretations, Wold's model concept is in the nature of a synthesis between a theoretical and an empirical content. The two model concepts are, however, complementary rather than conflicting; one difference is Bunge's asymmetry from theoretical to empirical notions, on the one hand, and Wold's symmetric treatment, on the other hand.

controlling knowledge already acquired concerning implications of the model structure.

It is of course chiefly results in the modelled real system of which, with the help of simulation, one tries to get some knowledge. It should thus be possible to get such knowledge through the interpretation of the model's structure. Modelling–simulation in interaction may consequently be seen as a hypothetico-deductive method; cf. Bunge (1967: I).

The above reasoning is compatible with the synthetic aspects of simulation advanced in section 1.1. In other words, simulation enters in the last phase in a procedure consisting of three phases: an analytic, a constructional and a synthetic phase. Summarily described, these three phases consist of the following elements:

(i) The analysis implies a breaking down of a complex problem concerning a "whole" into "parts" on one level, where earlier pronouncements and reasonable new hypotheses concerning the "parts" and their relations can be actualized and investigated.

(ii) The construction consists of the formulation of a model which implies a joining together of the "parts" and their relations which are obtained from the analysis.

(iii) The synthesis consists of simulation of "the behaviour of the parts" within the frame of the model.

The formulated model is thus assumed to be a complex system, and simulation is resorted to in order to obtain the sum total of the results of the "parts", i.e. in order to arrive at the behaviour of the "whole".

The main line in the discussion followed in the present work is derived from this application aspect of simulation. It may be pointed out that this view—i.e. that what is specifically new in the simulation method, more particularly in connection with the use of computers, is on the side of synthesis—has earlier been stated to be the most important aspect; for references see e.g. Abelson (1968).[1]

Many examples of the application of simulation for synthesizing purposes are reported in the literature. As an introduction to the study of different applications one may regard a survey worked out by Abelson (1968). It may be mentioned in passing that in economics this type of approach has been referred to as micro-analytical; an example of such a simulation study is that by Orcutt et al. (1961).

---

[1] In passing, also the possibility of using simulation in pure science for the evaluation of rival theories for the same empirical domain should be mentioned. As regards the use of formalized models (mathematics) in theory evaluation Bunge (1967: I, p. 476) remarks: "The region of disagreement of alternative theories can best be spotted, and the comparison of the respective virtues and faults can best be made if they are mathematized". Such a study may be undertaken, for example, by comparing simulation outcome from one model with the use of one theory after the other against a common set of empirical background material; cf. Zetterberg's notion and use of "gross predictions" of a theory (Zetterberg 1967).

2.5. Simulation and the posing of hypotheses

The simulation has been qualitatively changed and of course also improved by the advent and use of computers, which on the one hand have freed the method from manual calculation, and on the other hand afford richer possibilities for the creation of complex, dynamic and multi-variate models.

This implies that the model builder can shift his work from the deductive to the more demanding problems and, with respect to the alignment of aims, to directly dependent problems of an inductive kind. In the modelling work this consists of the formulation of hypotheses (the specification problem). This part of the modelling work—which in the last analysis seems referable to the transition from "facts" to theory—has always been a critical element. It is therefore probable that in the sequel this problem will even more be recognized as the central one, and will in course of time yield results in the form of a more developed theory for model construction than those now available.

This in part reorientation of model building, in which increased emphasis is laid upon the specification problem, marks a raising of aspiration levels. In particular, this transition weakens the emphasis on the other two fields of estimation and testing, which hitherto have been almost entirely dominant in theoretical as well as in applied statistics.[1]

2.6. Some first requirements and expectations of the study

We shall close this chapter by reviewing some of the requirements together with the consequent expectations that are considered important as regards the rationale of the present approach. By and large in line with R. A. Fisher's desiderata are the following requirements, which we class under the three heads: (i) Desiderata relating to subject-matter, (ii) Statistical desiderata and (iii) Formal desiderata.

(i) *Desiderata relating to subject-matter.* As in the present work the basis for modelling consists of a general model structure, this construction constitutes an important first step. The two perhaps most important requirements as regards an approach of this kind are:

- The structure should form the point of departure for the modelling of a class of social systems. This implies that it must be possible to provide

[1] The domination of the fields estimation and testing is primarily due to the impact of R. A. Fisher (1925, 1935), whose principles, however, were originally devised for experimental situations. Fisher's principles have, moreover, proved to have only limited scope in non-experimental statistics; cf. Wold (1956). For a frame of reference of, inter alia, experimental v. non-experimental means in research, reference is made to the flag table by Wold (1954, 1969*a*).

the structure with contents consisting of relations to the modelled systems represented.

- When it has been provided with a content the structure must be suitable both for a description of the modelled system and for solving problems, i.e. both the descriptive and operational aspects should be looked after.

The first requirement brings in its wake the question of the extent to which partial theories from the different social-scientific fields may be used. This part of the work pertaining to content includes, among other things, the fitting in of either or both of the following categories relating to statements:

- Statements that are already supported by empirical evidence but whose scope needs to be estimated or perhaps even controlled with reference to the special case.
- Statements in the form of hypotheses which refer to hitherto unknown or unexplored areas.

The work on the model structure thus requires a form that can successfully bind together the concepts in a network with connections that are already found and with new connections not yet established. It is assumed that the model structure that will be described offers the above possibilities. This gives us a glimpse of a way of carrying out synthesizing theoretical work consisting of combining partial theories from one scientific field or several such fields.

As for the empirical scope of the present modelling attempt, we shall follow the view held by Bunge that almost everything is to some extent explicable. Bunge goes on to consider defects and flaws in the search for explanations which prevent the finding of perfect or final explanations. These defects are due either to insufficient knowledge of laws and circumstances or to practical difficulties in the application of basically correct theories, or to both of these causes (Bunge 1967: II).[1]

(ii) *Statistical desiderata.* If the modelling work is to be performed with statistical methods then the model structure must offer the possibility of:

- giving an account, in connection with the specification of the content, of available knowledge of the modelled system and
- investigating the extent, in the treatment of the model, to which the deviations observed between modelling results and observed real results

[1] In other words, we are thus assuming that there is no basic limitation in the ability of the human mind to grasp even complex situations and in the penetration of these by models. Further, and this assumption seems more obvious, there is in principle no inherent limiting factor in the logic of the mathematics and statistics which will be used in the present modelling attempts. The reason for expressing these views is that there seems to be a parting of the ways here concerning the attitude towards the use of formalized models in the social sciences. We shall return to this discussion in section 3.3.

are reasonable, or ascertaining whether the deviations motivate a change of the model.

In the present approach this possibility is evidenced by the fact that the model structure will be given within a stochastic, in contradistinction to a deterministic, frame. This choice need not reflect any special philosophic position, but is rather to be regarded as the choosing of a feasible way of working.

(iii) *Formal desiderata.* To the statistical requirements of the approach should be added requirements of a purely logical nature which apply to all models. These are as follows:

- the model shall be free of contradiction in the sense that the premisses conflict with each other, and
- conclusions concerning the model must be obtained solely from deductions from the premisses; cf. Bunge (1967: I).

These requirements are practically met inasmuch as the model structure will be given in the form of a computer program and the conclusions are obtained through computer-processing of the model, in other words simulation on a computer.

Only a few aspects, of course, have been discussed, and this, so far, only in a summary way. For a more detailed treatment of the subject Bunge's criteria for theory evaluation provide a head start; see Bunge (1967: II). In Chapter 4 further general views of the rationale of the approach will be discussed from other points of departure after the introduction of the model structure.

Chapter 3

Problem areas in model building

3.1. Introduction

In this chapter five problem areas in connection with modelling are identified. Two of these areas, the preliminary construction and application of a developed model, are also briefly discussed.

In the next section the analysis is introduced with a general system-theoretical orientation. Then, in section 3.3, a presentation of different relations between the systems involved is given. Further studies will be devoted to the latter in the subsequent chapters. One of the main characteristics of the modelling approach is that for the most part the methodology follows cybernetic principles as presented by Klir & Valach (1967).

3.2. Systems involved in model building

In the literature that deals with the subject, the concept of systems has been described, and has been given verbal definitions, that to a greater or lesser extent deviate from the meaning intuitively understood. Briefly, the general meaning of the concept has been explained as implying a set of objects which are in some way related to each other, see e.g. Klir & Valach (1967). Apart from this general meaning there are, of course, special meanings given to the concept of systems. An example of this is the focussing on specific characteristics intended for the study of solutions, either for certain subject matters, or for certain types of problem. In such cases the situation is most often clarified by the use of specific prefixes, and one speaks of communication systems, education systems, control systems etc.

Owing to its universal applicability, the concept of systems is basic to the research aimed at here. This is because one assumes that the concept of systems, in the meaning of observations on relationships between existing parts, is the basis for the extending of knowledge about, and the description and explanation of, phenomena which at first are not directly viewed as interrelated.

The purpose of this work is, as has already been mentioned, to present a general model structure and its presumptive usage in the modelling of, in the main, social systems. The model structure is formulated in the symbology of mathematical statistics.

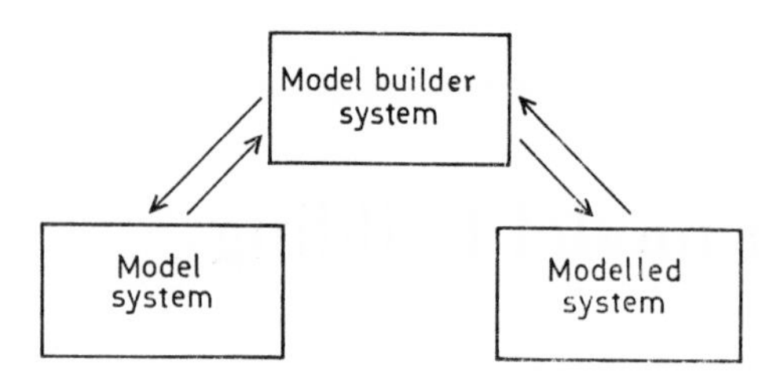

Fig. 3.1. The three subsystems in modelling.

In modelling, the concept of systems will be used as a superior concept. This means that this concept will also include the model builder himself and the modelled system, as well as mental creations like symbolically given model structures.

Figure 3.1 illustrates the interrelated subsystems in modelling described above. The model builder, besides considering the construction aspect, takes care of the registration of listed and revealed characteristics. This can be done for both the model system and the modelled system, with the purpose of controlling the perceived relationships between the two systems. The different relationships, with the model builder as the unifying link, are given a more precise treatment in the following chapters.

3.3. Problems concerned with model building

3.3.1. *On the use and application of models*

By way of introduction it is observed that the motive for using models which are built to illustrate complicated phenomena is—generally—to try to broaden the understanding of what one is dealing with, and to try to predict and give an insight into the probable outcomes in real systems. The chief motive for using parametric models in this connection is to seek in their quantitative alignment, which offers the possibility of statistical treatment.

According to the nature of the problem posed one can, opines Bunge, classify science as pure or applied science. If the problem is of a purely cognitive kind, the research in question is pure science. If, on the other hand, the nature of the problem is in the last analysis practical, and if the research is carried on with the same scientific methods as in pure science, then we have to do with applied science (Bunge 1967: I).

If the data about factual events yield the requisite attributes for the construction and testing of models in pure science, then the use of models for obtaining data concerning fictitious courses of events is the leading aspect of model building in applied science. Often one builds in efforts to control and, as far as possible, to change the "structure" and/or the "behaviour" of the model's real analogue.[1]

[1] Meanings of the concepts (hypothesis of the) structure and behaviour of a model (modelled) system will be given in chapter 4.

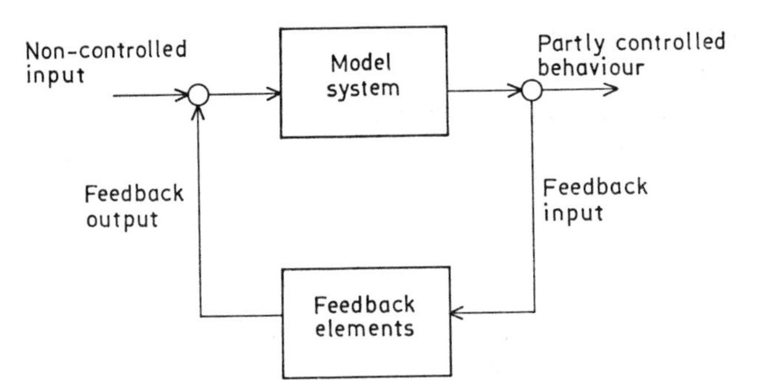

Fig. 3.2. The coupling of a model system with a feedback path.

In the simplest applied case, when some of the input-variables can be controlled in the real system, the results of the model can be optimized with respect to criteria that are expressed in terms of the desired behavioural development. In other words, the model is used for making conditional predictions giving information about how behaviour develops if one course of action is followed rather than another. A system, used as a model, can be incorporated into a larger feedback-control system like that shown in Fig. 3.2.

If no controllable variables exist the model will give forecasts from inputs that have been obtained exogenously. Forecasts are thus unconditional predictions and do not provide information about effects from alternative courses of action. In this case the model can be used in a larger feedback-control system, which contains the model as a subsystem.

The principles behind the use of models whose purpose is to direct and control are easily accessible. It should be mentioned that the so-called optimization theory contains a number of programming techniques for the maximization of criteria and value functions which reflect what one is trying to control, see e.g. Gue & Thomas (1968) for a survey of this field.

Apart from the above-adduced motives for the use of models in applied science, mention may be made of several others, inter alia, teaching and training, not least the training of the model-builder himself. As regards the use of—especially—computer models, I refer to the earlier mentioned survey by Abelson (1968) and to the introductory chapter by Dawson in Guetzkow (1962).

3.3.2. *On the construction of models*

Before a model can be regarded as giving an account of acquired knowledge (pure science) or before it is applicable (applied science), however, a considerable amount of work must be done, both on the initial construction of the model and on the subsequent adaptation of the model to the real system. These points will be dealt with specifically in the following chapters. Briefly, the study concerns the formation of pairs of systems; the modelling of real systems S^R with models S^M where

- S^R is a particular real system, of which the structure is not known and cannot be obtained explicitly and
- S^M on the one hand consists of a structure S given in the form of a computer programme and on the other hand consists of a content in the form of a model specification with references from the structure to the real system S^R.

One problem area in modelling thus deals with the first phase in the form of construction of a first version of the model system S^M.

The construction phase is often characterized as the most important and critical step in model building. Maybe the cautiousness in this assessment reflects the view that research in this area is too often carried on under the assumption that there exists one, and only one, possible model for a given real system, and this in spite of the fact that research work gives innumerable examples of cases where the latest findings give reasons for changing and supplementing existing models; cf. Bunge (1967: II). Because of this it seems more suitable, and more successful, to assume that there exists a set of parallel model systems belonging to the real system under study. The difficulty thus lies in the construction of a model that permits comparisons with existing accumulated knowledge in its area, yet also meets other demands, such as simplicity and economy of thought.[1]

The above reasoning is based on the observation that at the moment there is a relatively limited amount of knowledge concerning the relative merits of different approaches to the problem of model building vis-à-vis the social sciences.[2] To set up relationships between, for example, the different structures of model systems and their conduct in relation to modelled systems appears to be more a task for future theoreticians of model building. The research direction would then be a search for the acquirement of knowledge about, and preferably establish concordance and affinity between, model and modelled systems.

It is interesting to consider the construction problem in the light of a "natural law" from cybernetics which is known as "the law of requisite variety" (Ashby 1956). Briefly, this law says of this problem that every conception of a system sets an absolute limit to the variety of situations permitted by and comprehended in this conception. The problem of making

---

[1] As regards limitations in the construction of models, reference is made to section 2.6. In passing, we should also make reference to Alexander (1964) for his inspiring analysis of design problems, more especially against the background of more cautious expositions on related matters, e.g. by Popper (1961). However, it falls outside the scope of this work to deal any further with views on "possibilities" and their implications as regards research strategy.

[2] On the other hand, the argument by Kendall (1968), where the position is taken that no theory of model building in the social sciences may as yet be considered to exist, is perhaps too extreme.

a reasonable model may thus be discussed in terms of the variety that exists in the real system and the variety existing in the model.

Although wealth of variety in the individual parts of a real system is great, the total variation is fortunately not so great as may perhaps appear. This is due to the presence of organization of the parts, which imposes constraints and thus reduces the wealth of variation in the whole. From this cybernetic point of view the job of solving the specification problem is helped by the discovery of such constraints. This way of viewing the matter resembles statistical arguments, in which the aims are often formulated in terms of the study of regularities and stabilities in collectives.

The construction phase is described in Chapter 9, where model building concerning some teaching and learning processes is dealt with in relation to the application aspects in the previous subsection. This has been done mainly with the purpose of obtaining, if possible, extended knowledge, and being able to apply this knowledge to practical work in educational planning.

The revision of previously constructed models, when one has obtained a negative result from a confrontation with real systems also belongs to the construction problem.

In those cases where the existing models are too complex, or are otherwise uneconomical to use, then the constructional need to reduce the model will be clearly manifested. In other words, the problem concerns the simplicity of the models. These reduction problems lead to situations that are somewhat similar to the problem of constructing an original model.[1] In both cases the central question concerns the replacement of one system by another system which resembles the first from some points of view, but is at the same time easier to study. The reduction problem is, however, considerably easier, since the structure of the system is given, and the only task is to get together its most important qualities.

The problem of model reduction is described in the Appendix. Some attempts at reducing a complex model to simpler linear and non-linear relation systems are given there.

3.3.3. *An outline of problems concerned with model building*

Besides the construction and application aspects there exist the following three problem areas. Firstly, the model system S^M needs to be tested—if the properties of this system satisfy the criteria in a model specification. This problem implies in its turn that the two systems, the model system

[1] Compare with a related statement: "There is a very interesting theoretical problem here, as to what we mean when we say that one model is an approximation to another. Unless we can find some solution to this problem we might as well give up model-building altogether, for every model we write down is a simplification of real life." (Kendall 1961, p. 14).

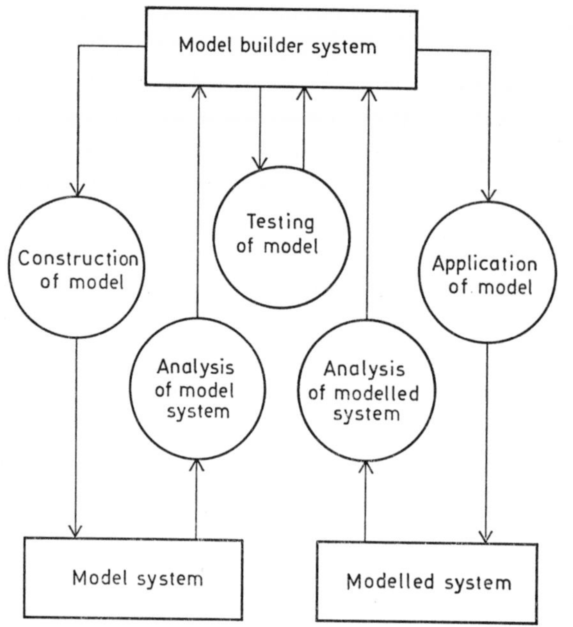

Fig. 3.3. The five problem areas in modelling.

S^M and the modelled system S^R, have been analysed and that knowledge about them has been acquired. Secondly, therefore, an analysis of the model system is required, an analysis of the behaviour of this system observed on the basis of the given and known structure. Thirdly, an analysis of the modelled system S^R is needed concerning the behaviour of the system, observed on the basis of hypotheses about this system in the form of the model S^M.

In this connection one can distinguish five problem areas. The areas, and their relationships with each other, are shown in Fig. 3.3, which is a development of Fig. 3.1. Naturally, model building aims at an alternating interaction between the reflected activities. In a real situation activities occur simultaneously in several of these areas.

Chapter 4
The simulation approach to modelling

4.1. Introduction

To be able to present the methods of model building which we have in view, one first of all needs a more detailed conception of the structure of the model, and an account of its most basic and characteristic qualities. In the following sections the model structure which we aim to use is formulated and discussed.

In the next section 4.2, some basic characteristics of the anticipated model structure are described, with, inter alia, the possibilities offered by the SIMULA language remembered.[1] Through this a picture of the intended model system's behaviour is also obtained. At the same time a frame of reference is obtained for the description of different phases of subsequent work with the model.

For pedagogical reasons the structure is presented as a generalization of the simple logistic system. After this it is shown that econometric relation systems are fairly well included in this framework. An implicit relation-representation of the systems' behaviour constitutes the next step. This is followed by a discussion of the reciprocal connection with attempts in the general systems theory, of which latter an account is given. Finally, the paradigm for simulation with the model is described.

In section 4.3 an account is given of the way in which observations coming from modelled systems are viewed and interpreted with the help of the model systems presented in the foregoing.

A specification of behaviour model is then given in section 4.4, together with indications of other possible specifications of models belonging to different levels of ambition.

4.2. The model systems

4.2.1. *Description of the structure*

When orienting oneself with the model structure a given starting point is in the present case the programming language that is used. This study is, as has already been mentioned, mainly based on the use of the SIMULA

[1] The SIMULA language is based on ALGOL and has been developed by Dahl & Nygaard (1965, 1966). See also subsequent works from the Norwegian Computer Center, Oslo, and the forthcoming publication by Birtwhistle et al. (1971).

language. A prerequisite for the understanding of the subject is, therefore, some understanding of SIMULA, or another similar language.[1] The structure is, accordingly, throughout assumed to consist of a computer programme. The exact form and structure of a model system is, in other words, found in a written computer programme. The consequences of a structure or "abstract theory" (section 2.3) are in their turn assumed to be obtainable by a running of the programme on a computer.

The structure is initially focussed with the logistic system as point of departure. With regard to some properties of the structure, the intended model structure constitutes a generalization of the logistic system. For this reason it appears convenient to introduce the present model structure as the end result from the removal of some of the limiting assumptions behind the logistic system.

The logistic system is in general used for the modelling of certain diffusion processes occurring in real systems.[2] The system may be described as consisting of a large number k of distinguishable subsystems $S_1. ..., S_k$. At each time-point τ, every one S_i of these systems takes on one of two possible values of a state variable η_τ^i, say 0 or 1. Every one S_i of these systems further interacts once at each integral time-point τ with another, randomly selected system S_j. If the selected system S_j has its state η_τ^j equal to 0, the state of the selecting system S_i does not change. If, on the other hand, η_τ^j is equal to 1, $\eta_{\tau+1}^i$ at time-point $\tau+1$ changes its value to 1 with probability α if the state η_τ^i of system S_i at the earlier time-point τ was equal to 0. The state 1 is, moreover, an absorbing state. This ends our description of the assumptions behind the logistic system.

In passing, it may be as well to add the remark that one attractive feature of the logistic system is its simplicity. The consequences of the assumptions are readily deduced by analytic means. Assume that the probability for a system being in state 1 at an integral time-point τ is the same for all systems and represent this probability by η_τ. The two probabilities for a particular system interacting with another system in state 0 or 1 at time-point τ are then the same for every system and are $1-\eta_\tau$ or η_τ. This descriptive account of the logistic system gives the following difference equation for the probability

$$\eta_{\tau+1} = \eta_\tau + \alpha\eta_\tau(1-\eta_\tau) \tag{4.2.1}$$

[1] Examples of comparable programming languages are SIMSCRIPT II (Kiviat et al. 1968) and GPSS (IBM 1966). For surveys on simulation programming languages, see Kiviat (1971) and Palme (1970).

[2] In general terms, the logistic system is a system whose time-series of "aggregated behaviour" is evolutive and is "s"-shaped, passing from one level to another. There are several variants. We have chosen to represent the system in close agreement with Karlsson (1958). Bunge (1967: I) uses the logistic system in his illustration of the applicability of mathematics in the formalization process. For other representations, see e.g. Bartlett (1960) and for brief surveys of representations of diffusion and growth process, see Törnquist (1967) and Coleman (1964); also for further references.

for a system in state 1 at the next integral time-point $\tau+1$. It also follows that the above probability may be interpreted as the expected proportion of all subsystems in state 1.[1]

Despite its simplicity, the starting point in the present search for a model structure is constituted by the interactive features of the logistic system. The main characteristics in this respect are that these interactions happen in time and take place between individual subsystems of the total system. It is thus assumed that these subsystems are separable from each other, which in a modelling context implies in turn that they may be located and identified in some specified space-time area. It should be pointed out that this question of localization has not been restricted to the common Euclidian space. As will appear later, it is the behavioural and functional aspects that will be considered the most important. There is therefore no reason to require that the structure should necessarily model the real system in its spatial organization.

While the logistic system has its field of application preferably in the modelling of simple diffusion processes occurring in some closed systems, it certainly has its limitations for the modelling of processes in social systems; cf. Coleman (1964). In particular, the limitations are decisive when the object is the modelling of interactive processes in which humans take active and integral parts. A consideration of the described assumptions for and restrictions on the logistic system reveals the following desired generalizations.

The structure should make it possible to allow for:

- different classes of subsystems, the members of one class being different from members of other classes as regards "structure" or "behaviour" or both,
- alternative formulations of the selection and state change "mechanisms", as well as extensions to other kinds of "mechanism",
- extension to vector state variables and incorporation of two new kinds of variable: vector input (stimulus) and output (response) variables,
- more flexible time sequencing of "events" than the unit step incrementation of a time variable taking on only integral values,
- extension to open systems implying interaction possibilities between an environment and the system.

---

[1] As will become clear later, we confine ourselves to processes related to discrete time. For the sake of completeness we give the difference equation (4.2.1) in continuous time as

$$d\eta/d\tau = \alpha\eta(1-\eta). \tag{4.2.2}$$

This differential equation has the solution

$$\eta = 1/(1+c\exp\{-\alpha\tau\}), \tag{4.2.3}$$

where the constant c is determined by the probability η_0 at some time point τ_0.

With these desiderata in mind we tentatively set out the following structure aimed at the modelling of dynamic interaction and behaviour in social systems. We call the structure a dynamic interaction simulator. The structure, in its general form, is represented as S. The system is presented in two different ways: on the resolution levels *zero* and *one* respectively. For a discussion of the concept resolution level of a system; see Klir & Valach (1967).

At the resolution level *zero* the system S consists of a single object A that interacts with its environment through input (stimulus) variables $\boldsymbol{\zeta}$ and output (response) variables $\boldsymbol{\xi}$. The system is dynamic, and its output depends not only on the most recent input, but also on previous inputs as well. These retained quantities are called the (internal) state $\boldsymbol{\eta}$ of the system; cf. Klir & Valach (1967).

The input and output variables change at discrete time-points. The value of the state variable is also changed at discrete time-points. The time-points at which changes occur are called events. In accordance with the design objectives of the SIMULA language, the system can likewise be called a discrete event system. The meaning of the behaviour of the system is made clear by the following definition.

Definition 4.1. *The behaviour of system S at the resolution level* zero *is given by the discrete time-series of state $\boldsymbol{\eta}$ and output $\boldsymbol{\xi}$ from an object A, which follows a given discrete time-series of applied inputs $\boldsymbol{\zeta}$ to the object A from the environment of S.*

Furthermore, the system is stochastic and shows genuine variability.[1] This means that the behaviour of the system can be viewed as realizations from probability distributions. Observations of the behaviour can consequently be viewed as the outcome of random drawings from these distributions.

The probability distributions together make up the structure of the system, and are called the system's mechanism; cf. Bråten (1968*b*). Other names are the system's operating characteristics (Orcutt et al. 1961) and operation rules (Dahl & Nygaard 1965). There has been no attempt to give any specific form for the mechanism of system S in the language of mathematical statistics. The mechanism is therefore written quite generally as $\mathbf{M}(\boldsymbol{\eta}, \boldsymbol{\xi} | \boldsymbol{\zeta}, \boldsymbol{\alpha})$, where $\boldsymbol{\alpha}$ is a mechanism parameter.

---

[1] A definition of genuine variability in the distinction genuine *v* apparent variability or scatter of a structure may be found in Wold (1969*b*). In a modelling context the genuine variability of a structure may model apparent variability of the real system in question, as is reflected by observational errors and/or genuine variability of the real system. As will later become clear, we have in the present case chosen to let the structure's genuine variability represent assumed genuine variability in the real system.

Example 4.1. A mechanism of the system at the resolution level *zero* can be exemplified by starting from the logistic system. This system is here viewed as a "whole" characterized by one single state variable η_τ representing the proportion of the k subsystems in state 1 at time-point τ. The mechanism giving the next state allows the formulation

$$\eta_{\tau+1} = \eta_\tau + \alpha\eta_\tau(1-\eta_\tau) + \varepsilon_{\tau+1} \tag{4.2.4}$$

with

$$E(\eta_{\tau+1} | \eta_\tau) = \eta_\tau + \alpha\eta_\tau(1-\eta_\tau), \tag{4.2.5}$$

implying that the disturbance $\varepsilon_{\tau+1}$ has mean equals zero.

Example 4.2. In a more general formulation the mechanism of the system can be exemplified by starting from the conditional frequency function

$$f(\boldsymbol{\eta}_{\tau+\Delta\tau}, \boldsymbol{\xi}_{\tau+\Delta\tau}, \Delta\tau | \boldsymbol{\eta}_\tau, \boldsymbol{\zeta}_\tau, \boldsymbol{\zeta}_{\tau'}, \tau, \tau', \boldsymbol{\alpha}), \tag{4.2.6}$$

where $\boldsymbol{\zeta}_\tau$ and $\boldsymbol{\zeta}_{\tau'}$, are two consecutive inputs that are applied to the system at time-points τ and τ', where $\tau < \tau'$.

An observation of the behaviour may thus be seen as consisting of a random drawing from the conditional distribution above. The system handles a sequence of inputs sequentially. After a realization the conduct follows the new rule that follows from the formulation (4.2.6).

Starting from the conception of a mechanism, the following is given.

Definition 4.2. *The structure of the system S at resolution level* zero *consists of a mechanism $\mathbf{M}(\boldsymbol{\eta}, \boldsymbol{\xi} | \boldsymbol{\zeta}, \boldsymbol{\alpha})$, which is associated with the object A. The mechanism determines the behaviour of the system.*

The behaviour and structure can be illustrated in the way shown in Fig. 4.1.

At resolution level *one* system S is seen as consisting of a group of k distinguishable objects, $A_1, ..., A_k$. Every one of these objects interacts with every other, and also with the environment to system S.

If , for every object A_i, one lets (sub) system S_i be defined analogously to system S at resolution level *zero,* then each system S_i is a discrete event system, with a structure like the one of system S at resolution level *zero.* For system S_i let the state be $\boldsymbol{\eta}^i$, the input $\boldsymbol{\zeta}^i$ and the output $\boldsymbol{\xi}^i$, while the mechanism is written as $\mathbf{M}_i(\boldsymbol{\eta}^i, \boldsymbol{\xi}^i | \boldsymbol{\zeta}^i, \boldsymbol{\alpha}^i)$.

Since the environment to every subsystem S_i now consists of several objects, the inputs and outputs for different subsystems must now be distinguishable and directed to the correct subsystem(s). The way in

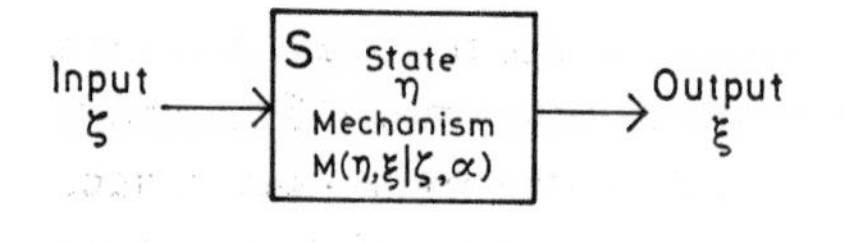

Fig. 4.1. Illustration of the behaviour and structure of the system S at resolution level *zero*.

which inputs and outputs for the different systems are connected can be formulated in a convenient way if these inputs and outputs are written

$$\zeta^i = \begin{pmatrix} {}^0\zeta^i \\ {}^1\zeta^i \\ \vdots \\ {}^k\zeta^i \end{pmatrix} \qquad \xi^i = \begin{pmatrix} {}^0\xi^i \\ {}^1\xi^i \\ \vdots \\ {}^k\xi^i \end{pmatrix}, \tag{4.2.7}$$

where ${}^j\zeta^i({}^j\xi^i)$ is the input to (the output from) system S_i from (to) the system S_j, and where, for the sake of simplicity, the common environment to all subsystems has been written S_0. If ${}^j\zeta^i = 0({}^j\xi^i = 0)$, this means that no such input (output) exists.

The system is completely determined if the following fundamental equations hold,

$${}^j\zeta^i = {}^i\xi^j \quad i, j = 1, \ldots, k \tag{4.2.8}$$

that is, the input to the system S_i from the system S_j is equal to the output from the system S_j to the system S_i.

Example 4.3. The set of k mechanisms of the system at resolution level *one* can be exemplified by starting from the logistic system as it was originally described above. The system was then viewed as consisting of k (sub)systems $S_1, \ldots, S_k$. Every one system S_i was further characterized by a single state variable η^i. For every system S_i, the associated mechanism $\mathbf{M}_i$ can be given the approximative formulation

$$\eta^i_{\tau+1} = \begin{cases} 1 \text{ if } \eta^i_\tau = 1 \text{ or with probability } \alpha \text{ if } \eta^i_\tau = 0 \text{ and} \\ \quad \zeta^i_\tau \text{ contains at least one element equal to 1} \\ 0 \text{ otherwise} \end{cases} \tag{4.2.9}$$

$${}^j\xi^i_{\tau+1} = \begin{cases} \eta^i_{\tau+1} \text{ with probability } \dfrac{1}{k-1} \\ \\ 0 \text{ otherwise.}^1 \end{cases} \qquad \text{for } j = 1, \ldots, i-1, i+1, \ldots, k \tag{4.2.10}$$

[1] In the strict sense, the given assumptions for the logistic system imply that the input ζ^i consists of zeros at all places except for the one giving the state of the "selected" system.

Example 4.4. A more general formulation of the set of k mechanisms, associated with the k subsystems $S_1, \ldots, S_k$ respectively, can be obtained by starting from the formulation (4.2.6). Thus, for every subsystem S_i, the mechanism can be exemplified by

$$f_i(\boldsymbol{\eta}^i_{\tau+\Delta\tau_i}, \boldsymbol{\xi}^i_{\tau+\Delta\tau_i}, \Delta\tau_i \,|\, \boldsymbol{\eta}^i_\tau, \boldsymbol{\zeta}^i_\tau, \boldsymbol{\zeta}^i_{\tau'}, \tau, \tau', \boldsymbol{\alpha}^i) \tag{4.2.11}$$

with an interpretation similar to that in example 4.2.

The behaviour and structure are defined as

Definition 4.3. *For a given time-series of applied inputs to the objects from the environment of S,*

$$^0\boldsymbol{\zeta} = \begin{pmatrix} ^0\boldsymbol{\zeta}^1 \\ \vdots \\ ^0\boldsymbol{\zeta}^k \end{pmatrix} \tag{4.2.12}$$

the behaviour of the system S at resolution level one *is given by the time-series of states, inputs and outputs*

$$\boldsymbol{\eta} = \begin{pmatrix} \boldsymbol{\eta}^1 \\ \vdots \\ \boldsymbol{\eta}^k \end{pmatrix} \quad \boldsymbol{\zeta} = \begin{pmatrix} \boldsymbol{\zeta}^1 \\ \vdots \\ \boldsymbol{\zeta}^k \end{pmatrix} \quad \boldsymbol{\xi} = \begin{pmatrix} \boldsymbol{\xi}^1 \\ \vdots \\ \boldsymbol{\xi}^k \end{pmatrix} \tag{4.2.13}$$

for the objects $A_1, \ldots, A_k$ respectively.

Definition 4.4. *The structure of the system S at resolution level* one *is defined by the population of mechanisms*

$$\mathbf{M}_i(\boldsymbol{\eta}^i, \boldsymbol{\xi}^i \,|\, \boldsymbol{\zeta}^i, \boldsymbol{\alpha}^i) \quad i = 1, \ldots, k \tag{4.2.14}$$

that are associated with the objects $A_1, \ldots, A_k$ respectively. The mechanisms determine the behaviour of the system.

The behaviour and structure of system S at resolution level *one* is illustrated in Fig. 4.2.

The basic features of the abstract system S have thus been described. It is, of course, possible to increase the formal representation of the system. An example which shows an important structural aspect is given in the work by Klir & Valach (1967). Consider, for example, inputs and outputs from system S when the system is given at resolution level *one*. Define the structural parameters

$$\omega_{ij} = \begin{cases} 0 \text{ if } {}^j\boldsymbol{\zeta}^i \text{ is always equal to zero} \\ 1 \text{ otherwise.} \end{cases} \tag{4.2.15}$$

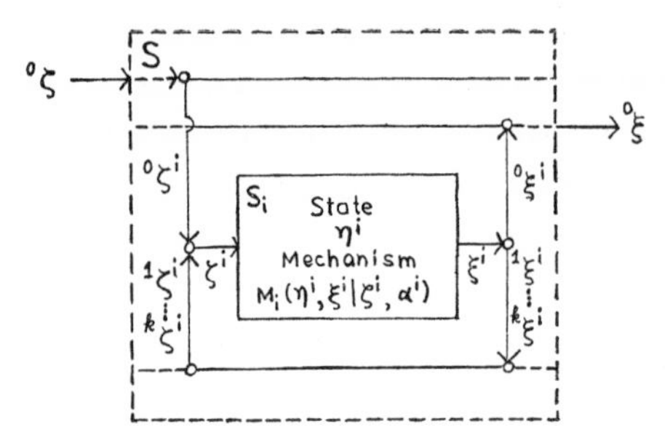

Fig. 4.2. Illustration of the behaviour and structure of the system S at resolution level *one*.

A structural aspect of system S at resolution level *one* can thereby be given by the elements off the main diagonal in the square "coarse structure" matrix

$$\mathbf{\Omega} = (\omega_{ij}), \quad i, j = 0, 1, \ldots, k; \quad i \neq j \tag{4.2.16}$$

of the order $k+1$, where ω_{ij} is defined by (4.2.15).[1]

In the same way the resolution level can be raised even more. At resolution level *two*, therefore, every system S_i consists of a set of k_i distinguishable objects A_{ij}, $j = 1, \ldots, k_i$, where every object A_{ij} is supplied with a mechanism $\mathbf{M}_{ij}(\boldsymbol{\eta}^{ij}, \boldsymbol{\xi}^{ij} | \boldsymbol{\zeta}^{ij}, \boldsymbol{\alpha}^{ij})$ and identified as a subsystem etc.

Even if the resolution level is a central concept, it is still relative and can only be made meaningful in a given context. This follows from the fact that systems are presented differently if they are observed from different viewpoints. A definition that can be used logically, however, is given by

Definition 4.5. *The resolution level of a description of the system S is equal to the greatest number of levels of subsystems in the hierarchical building up of S as implied by that description, and as distinct from another description of S implying a smaller or larger number of levels of subsystems of S.* (This usage is consistent with Bråten (1970 *a–b*).)

4.2.2. *Representation of the structure in the SIMULA language*

Principally, the system-objects A are found in the computer programme as separate programme blocks. All objects having the same structure, however, need only be described once. Starting from this description, several objects belonging to the same class can be generated when the programme is being run. Naturally, it is possible in this connection to give different values to the initial states of different objects.

In these circumstances the class concept of the SIMULA language stands out as a central idea in the description of the objects. This class concept allows a hierarchical building up of the objects at some resolution level. This means that a subclass of objects can be described as a projection from a class all objects belong to. The class then has all the properties of the

[1] Reference is made to the seemingly related though not identical concept precedence matrix; see Langefors (1966).

more general class, and also those properties that have been found significant for this subclass. In terms of the system S at resolution level *one* this can be illustrated as follows. Those parts of the mechanisms $\mathbf{M}_i$ of the objects A_i that perform the same operations can be described jointly in a separate programme block.

A consequence that follows from the formulation of system S at higher resolution levels is that the objects are presumed to exist parallel to each other in time. At every time-point several objects may be in interaction phases with their environments, while the state of other objects may be changing, etc. The computer works sequentially, and this implies that the programming language must in some way allow the possibility of treating the dynamic and simultaneously existing subsystems in a quasi-parallel way. This possibility is found in SIMULA. The pseudo-parallel running possibilities in this language mean that the computer alternately calculates through one of the programme blocks, and then through another, as the simulated time is moved along a constructed time-axis. In this way a course of events is given in which, among other things, interactions and changes of states occur. In other words, the behaviour of the system is given.

4.2.3. *Interpretations of the structure in terms of relations between variables*

The logical structure and form of system S that were described in a preceding subsection constitute the basis of modelling in this monograph. It would therefore be of value to try and get an idea of the applicability of the structure S already at this stage. The purpose of this section is to try to reach a first assessment of this kind.

The problem will be tackled by studying some structures that are used in econometrics. A partial and positive answer can, therefore, be given in an indirect way to the question set concerning the usability of system S. This answer refers to the possibility of showing that the structure of system S includes the proved structure of the econometric system.

The study deals with the special case where the mechanism (mechanisms) of system S consists (consist) of conditional distributions (4.2.6) and (4.2.11), as in the examples 4.2 and 4.4 respectively.

The observed structures in econometrics, as well as the structures in some other areas, are based on a linear system of relations that can be written as

$$\mathsf{B}\boldsymbol{\eta} = \boldsymbol{\Gamma}\boldsymbol{\zeta} + \boldsymbol{\delta}, \tag{4.2.17}$$

where $\boldsymbol{\eta}$, $\boldsymbol{\zeta}$ and $\boldsymbol{\delta}$ are stochastic vector variables, and B and $\boldsymbol{\Gamma}$ are two coefficient matrices. If the matrix B is non-singular the system can be written in the form

$$\boldsymbol{\eta} = \boldsymbol{\Omega}\boldsymbol{\zeta} + \boldsymbol{\epsilon}, \tag{4.2.18}$$

where

$$\begin{cases} \boldsymbol{\Omega} = \mathsf{B}^{-1}\boldsymbol{\Gamma} \\ \boldsymbol{\epsilon} = \mathsf{B}^{-1}\boldsymbol{\delta}. \end{cases} \tag{4.2.19}$$

In econometric terminology $\boldsymbol{\eta}$ is the endogenous and $\boldsymbol{\zeta}$ the predetermined variable. The variable $\boldsymbol{\epsilon}$ is an unobservable disturbance. The two variants (4.2.17) and (4.2.18) are called the structural and reduced forms respectively.[1]

The system will be looked at primarily from formal and statistical points of view. In general, the statistical properties of the system are specified by giving the joint distribution of (distribution characteristics for) the $\boldsymbol{\zeta}$ and $\boldsymbol{\epsilon}$ variables. The distribution of (distribution characteristics for) the endogenous variable $\boldsymbol{\eta}$ can then be obtained from (4.2.18). The problems concerned with this are discussed in econometrics. The debate is not focussed on the use of the linear model as such. Instead, the problem concerns what assumptions must be made for the joint distribution of the $\boldsymbol{\zeta}$ and $\boldsymbol{\epsilon}$ variables. The assumptions made may involve consequences for the estimation of the coefficient matrices $\mathbf{B}$ and $\boldsymbol{\Gamma}$, and for the interpretation and operative use of these estimates.

(i) *The reduced form.* First the reduced form (4.2.18) is observed. One usually introduces the time-dimension in econometrics by letting the interval between subsequent observations be one time-unit. The system can then be written

$$\boldsymbol{\eta}_{\tau+1} = \boldsymbol{\Omega}\boldsymbol{\zeta}_{\tau} + \boldsymbol{\epsilon}_{\tau+1}, \tag{4.2.20}$$

where the time-variable τ takes integral values. It is very likely that this system will be regarded analogously to system S at resolution level *zero*. If one lets the endogenous variable $\boldsymbol{\eta}_{\tau+1}$ be the common state and output variable of the object A from an input $\boldsymbol{\zeta}_{\tau}$ which is applied to the system, then the following characteristics of system S can be noted: (i) fixed time-intervals between inputs, (ii) fixed time-intervals between one input and its corresponding output and (iii) the time invariant structure.

The conditional frequency function (4.2.6) then takes the following simple form

$$f(\boldsymbol{\eta}_{\tau+1} | \boldsymbol{\zeta}_{\tau}, \boldsymbol{\alpha}). \tag{4.2.21}$$

When one compares (4.2.20) and (4.2.21) the following possibility emerges[2]

$$E(\boldsymbol{\eta}_{\tau+1} | \boldsymbol{\zeta}_{\tau}, \boldsymbol{\alpha}) = \boldsymbol{\Omega}\boldsymbol{\zeta}_{\tau}. \tag{4.2.22}$$

---

[1] The terminology and the classical assumptions that will be made can be found in any standard textbook in econometrics, see e.g. Johnston (1963), Goldberger (1964) or Malinvaud (1964).

[2] Throughout this work we shall make extensive use of conditional expectations. The notion and use of predictors or as they are also called, *eo ipso* predictors as synonymous with stochastic relationships specified by conditional expectations are associated with the name H. Wold. He argues for the use of predictors in the specification of model mechanisms serving purposes of prediction or causal analysis as in the specification (4.2.27); see Wold (1959*b*) and subsequent works by him.

A consequence of the specification (4.2.22) as a conditional expectation is the often occurring classical specification of the system. It can be shown that the following properties of the disturbance can be derived from (4.2.22)

$$E(\boldsymbol{\epsilon}_{\tau+1}) = \mathbf{0} \tag{4.2.23}$$

$$E(\boldsymbol{\epsilon}_{\tau+1}\boldsymbol{\zeta}'_{\tau}) = \mathbf{0}. \tag{4.2.24}$$

The parameter $\boldsymbol{\alpha}$ in (4.2.22) contains the coefficients in the reduced form and also those parameters that determine the distribution of the disturbance term $\boldsymbol{\epsilon}$. It has thus been shown that it is possible to interpret the econometric reduced form as a system S on a low resolution level.

(ii) *The structural form.* If the interest is concentrated on the structural form (4.2.17), then we may try to interpret this specification in terms of system S at resolution level *one*. Analogously to (4.2.20) the time-dimension for the structural form is introduced in the following way,

$$\mathsf{B}\boldsymbol{\eta}_{\tau+1} = \boldsymbol{\Gamma}\boldsymbol{\zeta}_{\tau} + \boldsymbol{\delta}_{\tau+1}. \tag{4.2.25}$$

The course of action connecting the econometric system with the system S follows two lines of development.

First attempt. The first method is to try to derive the mechanisms (4.2.11) from (4.2.25) direct. If the definition of system S is studied and taken as the starting point, we may interpret the system as consisting of n objects, one object for each relation. Furthermore, one lets the endogenous variable η_i be the common state and output variable associated to subsystem S_i and object A_i. If one adopts a suitable numbering it is possible to let the ith relation contain endogenous variable η_i for $i = 1, \ldots, n$.

The way of attacking the problem is hence to investigate under which circumstances the relationships can be treated autonomously as the basis of the mechanism specifications. The input variables of the system S_i in this case consist of those endogenous variables η_j (with $j \neq i$) and those predetermined variables that actually appear in the ith relation. One can represent these two sets of variables as $\boldsymbol{\eta}_{(i)}$ and $\boldsymbol{\zeta}_{(i)}$ respectively. The conditional distribution of the state variable of subsystem S_i should hence have the form

$$f(\eta_{i\tau+1} \mid \boldsymbol{\eta}_{(i)\tau+1}, \boldsymbol{\zeta}_{(i)\tau+1}, \boldsymbol{\alpha}^i) \tag{4.2.26}$$

If we look at the above expression, taking into account the linear form of the ith relation, then the following appears possible

$$E(\eta_{i\tau+1} \mid \boldsymbol{\eta}_{(i)\tau+1}, \boldsymbol{\zeta}_{(i)\tau}, \boldsymbol{\alpha}^i) = -\boldsymbol{\beta}'_{(i)}\boldsymbol{\eta}_{(i)\tau+1} + \boldsymbol{\gamma}'_{(i)}\boldsymbol{\zeta}_{(i)\tau}, \tag{4.2.27}$$

where $\boldsymbol{\beta}_{(i)}$ and $\boldsymbol{\gamma}_{(i)}$ are parameter vectors which contain the non-zero elements from the ith row of the matrices B and $\boldsymbol{\Gamma}$, and which correspond to the variables $\boldsymbol{\eta}_{(i)}$ and $\boldsymbol{\zeta}_{(i)}$. It is known that the specification (4.2.27) is in general limited to special econometric systems; that is, those systems that are known as vector regression (VR) and causal chain (CC) systems. Furthermore, the specification of the system in terms of a set of n autonomous mechanisms demands that the disturbance terms in $\boldsymbol{\delta}_{\tau+1}$ in (4.2.25) be independently distributed. Among other things, this in its turn implies that the disturbances are uncorrelated.

It is of some interest to notice the similarity between these conclusions and some viewpoints that were introduced into econometrics in the 60's. A key question in the debate at this time was the way in which the stochastic framework ought to be applied to the structural form. It was maintained that interdependent (ID) systems were from a statistical point of view problematical. The use of VR- and CC-systems was recommended as far as possible on account of their more interpretable forms.[1]

The present results support this argument, since it has been shown that the system-theoretical orientation in terms of system S cannot, for econometric systems, be extended so as to be valid for ID-systems, but only for VR- and CC-systems.

Second attempt. The limitation to VR- and CC-systems can be removed if one takes into account a reformulation of the structural form.[2] The classical system (4.2.25) is hence reformulated as

$$\boldsymbol{\eta}_{\tau+1} = \mathsf{B}^*\boldsymbol{\eta}^*_{\tau+1} + \boldsymbol{\Gamma}\boldsymbol{\zeta}_\tau + \boldsymbol{\epsilon}_{\tau+1}, \tag{4.2.28}$$

where the coefficients are numerically the same as before, that is

$$\mathsf{B}^* = \mathbf{I} - \mathsf{B}. \tag{4.2.29}$$

The variable $\boldsymbol{\eta}^*$ is defined as the systematic part of the reduced form. In symbols

$$\boldsymbol{\eta}^*_{\tau+1} = \boldsymbol{\Omega}\boldsymbol{\zeta}_\tau. \tag{4.2.30}$$

This reformulation means that the structural form has the same reduced form as before. Moreover, the disturbances will now be the same in both the structural and reduced forms, as is indicated in (4.2.28).

The interpretation is made in the following way. The state variables for subsystem S_i are η_i and η^*_i. The input variable for the other subsystems is η^*_i and the output to the environment is η_i. Consider an input $\boldsymbol{\zeta}_\tau$ applied to the system S at time-point τ. It is by this means possible to create an internal and deterministic input/output sequence which gives

$$\boldsymbol{\eta}^*_{\tau+k/K} = \mathsf{B}^*\boldsymbol{\eta}^*_{\tau+(k-1)/K} + \boldsymbol{\Gamma}\boldsymbol{\zeta}_\tau;\ k=1,\ \ldots,\ K \tag{4.2.31}$$

[1] VR-systems can be categorized as a special case of CC-systems. The distinction between CC- and ID-systems seems first to have been made by Bentzel & Wold (1946). Until the ID-systems approach was made clear, the CC-approach was advocated in particular by Wold; see Wold (1964) and Meissner (1971).

[2] The reformulation is due to Wold (1959*b*, 1960) and made possible through his systematic utilization of predictors (footnote 2 p. 38 above).

If the characteristic roots of the matrix B^* are numerically smaller than one, the input/output sequence gives the final values

$$\boldsymbol{\eta}^*_{\tau+1} = (\mathbf{I} - \mathsf{B}^*)^{-1}\boldsymbol{\Gamma}\boldsymbol{\zeta}_\tau = \mathsf{B}^{-1}\boldsymbol{\Gamma}\boldsymbol{\zeta}_\tau = \boldsymbol{\Omega}\boldsymbol{\zeta}_\tau, \tag{4.2.32}$$

to any prescribed accuracy, if the number of steps K in the sequence is large enough. In the special case where the system is a VR-(CC-)system it is sufficient if K equals $1(n-1)$.

The behaviour of subsystem S_i at time-point $\tau+1$ will then depend on the conditional distribution

$$f_i(\eta_{i\tau+1} \mid \eta^*_{i\,\tau+1}, \boldsymbol{\alpha}^i), \tag{4.2.33}$$

since the endogenous variable can be written as

$$\eta_{i\tau+1} = \eta^*_{i\,\tau+1} + \varepsilon_{i\tau+1}. \tag{4.2.34}$$

Even in this interpretation the specification of the system demands that the disturbances be distributed independently of each other and of the input, that is the predetermined variables.[1] Specifically, (4.2.24) is still valid.

It is recognized that there exist certain statistical distribution limitations in the interpretation of econometric systems within the framework of system S. Where the necessary restrictions differ from the specifications usually employed, one should take into account the fact that econometric specifications have other starting points, the estimation problem being one of the most often emphasized in this respect. There are, however, reasons to observe that the specifications are to a large extent the same. As a secondary result, the analysis gives new insights into the representations (4.2.17–18) of an econometric system. The reduced form can be interpreted at resolution level *zero*, while the structural form can be interpreted at resolution level *one*.

When summarizing the results of this section we discover that the range of system S extends tolerably well over the whole range of econometric systems. The reverse, however, seems not to be valid.

In modelling with relation systems there are only implicit correspondences between the objects of the model system and the objects of the modelled system. Furthermore, it is difficult to keep these model correspondences in the analytical and numerical treatment. Some fundamental forms concerning the organization of the objects are lost in obtaining an abstract variable-system, which in the formulation may be said to be deprived of its physical identity. For another treatment of the distinction between relation systems and other, more general, systems, reference is made to Angyal (1941).

[1] There are other, as well as similar, interpretations of ID-systems reported in the literature, e.g. that by Strotz & Wold (1960), describing how an ID-system may be seen as a limiting approximation to a VR- or CC-system in which certain time-lags approach zero. See also F. M. Fisher (1967) referred to by Theil (1971).

4.2.4. *Representation of the behaviour in terms of relations between variables*

At the end of the previous section the standpoint was taken that it is in general impossible to express the structure of system S as a form that is consistent with one or more relations between variables. This is a disadvantage for analytical work, and for an understanding of the structure. In the present studies this problem, relating to certain aspects of the state of system S, has been solved, at least to some extent. This can be done through the derivation of certain implicit relation representations involving the behaviour of the system.

Observe the structure of system S. The mechanism parameters are written as $\boldsymbol{\alpha}$. The vector of all state variables at time-point τ is written as $\boldsymbol{\eta}_\tau$.

If a time-interval Δ is chosen small enough, then, in general, the system receives at the most one input from the environment during each time-period $(\tau, \tau+\Delta)$.

Choose a new scale for the time-variable so that the length Δ on the old scale corresponds to one time-unit on the new scale. For the sake of simplicity the old notation τ for the time-variable is retained. Characterize the input from the environment of the system during the time-period $(\tau, \tau+1)$ by ${}^0\boldsymbol{\zeta}_\tau$.

The following relation system for the state of the system is considered valid with a sufficient degree of accuracy

$$\boldsymbol{\eta}_{\tau+1} = \boldsymbol{\varkappa}(\boldsymbol{\alpha}, \boldsymbol{\eta}_\tau, {}^0\boldsymbol{\zeta}_\tau) + \boldsymbol{\delta}_{\tau+1}(\boldsymbol{\alpha}, \boldsymbol{\eta}_\tau, {}^0\boldsymbol{\zeta}_\tau), \tag{4.2.35}$$

where $\boldsymbol{\varkappa}$ is a vector function, and where the disturbance $\boldsymbol{\delta}_{\tau+1}$ is a stochastic vector variable that is defined by

$$E(\boldsymbol{\eta}_{\tau+1} | \boldsymbol{\alpha}, \boldsymbol{\eta}_\tau, {}^0\boldsymbol{\zeta}_\tau) = \boldsymbol{\varkappa}(\boldsymbol{\alpha}, \boldsymbol{\eta}_\tau, {}^0\boldsymbol{\zeta}_\tau). \tag{4.2.36}$$

The time-series of the state starts with the initial value $\boldsymbol{\eta}_0$ at time-point zero, and with given $\boldsymbol{\alpha}$ and ${}^0\boldsymbol{\zeta}_\tau$. We shall in what follows refer to $\boldsymbol{\eta}_0$ and the time-series of ${}^0\boldsymbol{\zeta}_\tau$ as predetermined variables. Let $\boldsymbol{\eta}$ represent a population of state variables, at a population of observed time-points. From this it follows that the state $\boldsymbol{\eta}$ can be written as

$$\boldsymbol{\eta} = \boldsymbol{\phi}(\boldsymbol{\alpha}, \boldsymbol{\eta}_0, \boldsymbol{\zeta}) + \boldsymbol{\epsilon}(\boldsymbol{\alpha}, \boldsymbol{\eta}_0, \boldsymbol{\zeta}), \tag{4.2.37}$$

with

$$E(\boldsymbol{\eta} | \boldsymbol{\alpha}, \boldsymbol{\eta}_0, \boldsymbol{\zeta}) = \boldsymbol{\phi}(\boldsymbol{\alpha}, \boldsymbol{\eta}_0, \boldsymbol{\zeta}), \tag{4.2.38}$$

where $\boldsymbol{\phi}$ is a vector function, $\boldsymbol{\epsilon}$ is the disturbance and where $\boldsymbol{\zeta}$ is the vector of all inputs from the environment during the observed time-period. Although the characteristics of system S do not, in general, allow the above relationships to be derived explicitly, these expressions come to appear suitable for the description of certain operations on the system in what follows.

4.2.5. *Connections to general systems theory*

In the wide field that has come to be called general systems theory, the development of abstract and formalized systems is of special interest to this study. This interest goes hand in hand with the statistician's general need to incorporate quite general model structures with a wide scope.

The first points of view and ideas in the field of general systems theory are usually ascribed to L. von Bertalanffy (1945). Proceeding principally from biology, he designed an empirical and experimental method which has been summarized as follows: "System is considered as a complex physical object consisting of several parts, each of which is associated with some quantities that are in a relation to the quantities of the other parts. Applying this approach, systems belonging to different scientific disciplines are investigated in their natural forms. On the basis of experimental results, isomorphic relations between different systems are studied and, finally, some general principles applicable for all systems of a certain class are formulated." [Klir (1969, p. 97), in a survey of the works by Bertalanffy.]

The meaning of this quotation points in a direction which, if extended, seems to be aimed at abstractions that have a form closely corresponding to that of system S.

Principally on the basis of cybernetics, Ashby has formulated and developed the "state determined system". First he describes an isolated system that is characterized by constant frame conditions and inputs. In a discrete form this system can be represented in the following canonical form,

$$\boldsymbol{\eta}_{\tau+1} = \boldsymbol{\phi}(\boldsymbol{\eta}_\tau), \tag{4.2.39}$$

where the vector variable $\boldsymbol{\eta}_\tau$ is the state of the system at time-point τ (Ashby 1952). In the discussion of this closed system it has been said that this form of description represents an ideal knowledge since, among other things, predictions from (4.2.39) are unambiguous and easily derived.

Starting from (4.2.39), it is interesting to look at the problem of "equifinality" that is discussed by Bertalanffy, amongst others. This concerns the independence of the initial conditions for the possible final state of equilibrium.[1] Since the system (4.2.39) might very well not be equifinable, Bertalanffy points out that real systems often exhibit an actual equilibrium that has originated from different initial conditions.

[1] I.e. the possible solutions of $\boldsymbol{\eta}^* = \boldsymbol{\phi}(\boldsymbol{\eta}^*)$. It follows from Kolmogorov & Fomin's theorem (1957) on "contraction mappings" that the equifinality of (4.2.39) exists if, together with continuity restrictions, the inequality $\|\partial\boldsymbol{\phi}/\partial\boldsymbol{\eta}\| \leqslant \delta < 1$ is valid.

The main reason for this is said to be that real systems are open, and are continuously influenced by each other (Bertalanffy 1950).

In the extension of (4.2.39) Ashby has shown and developed an open state determined system, which in discrete form can be formulated as

$$\boldsymbol{\eta}_{\tau+1} = \boldsymbol{\phi}(\boldsymbol{\eta}_\tau, \boldsymbol{\zeta}_\tau), \tag{4.2.40}$$

cf. (4.2.30), where $\boldsymbol{\zeta}_\tau$ represents the external conditions of the system at time-point τ.

From the system (4.2.40) it is not far, for example, to a system which Klir & Valach calls a "determinate system with sequential behaviour", which also serves as a representation of open systems. This system is a discrete and deterministic variable system which can be formulated as

$$\begin{cases} \boldsymbol{\eta}_{\tau+1} = \boldsymbol{\phi}_1(\boldsymbol{\eta}_\tau, \boldsymbol{\zeta}_\tau) \\ \boldsymbol{\xi}_\tau = \boldsymbol{\phi}_2(\boldsymbol{\eta}_\tau, \boldsymbol{\zeta}_\tau), \end{cases} \tag{4.2.41}$$

where $\boldsymbol{\eta}_\tau$, $\boldsymbol{\xi}_\tau$ and $\boldsymbol{\zeta}_\tau$ are the internal state, response and stimulus of the system at time-point τ; cf. (4.2.6). It should be pointed out, however, that we have here put the time-lag which Klir & Valach calls "reaction time" equal to one time-unit. Thus the change of state and response are unambiguously determined by the state at the earlier time-points, and by received stimuli (Klir & Valach 1967).

The above points illustrate some unifying features of the present approach with general systems theory. In this connection it may be emphasized that this theory has advanced a long way from the introductory and general conception of the systems concept. A large volume of literature has appeared in the course of the years. Outlines of the work in the various fields concerned are available in Klir (1969), amongst others.

To what extent the "ideal forms of knowledge" in general systems theory are useful when modelling in the social sciences is a question of great interest; the more so when one considers the, in many ways, limited knowledge available about these systems. This has been observed in cybernetic quarters: "... Because knowledge in this form is complete and maximal, all the other branches of the theory, which treat of what happens in other cases, must be obtainable from this central case as variations on the question: what if my knowledge is incomplete in the following way ...?" (Ashby 1966 (1952), p. 270). Ashby has systematically investigated the problems arising when, for example, the form (4.2.41) is incompletely known.

It appears an important task, therefore, to develop this method further with, among other things, structures that explicitly take into account the fact that certain modelled elements are incompletely known. One way to develop the method is, as is done in this work, to see the incoming variables from a stochastic viewpoint as opposed to a deterministic one. Therefore an important achievement in general systems theory seems to be the extension to, for example, a system where the relations (4.2.41) are extended so as to incorporate stochastic variables as well as disturbance terms, which Klir & Valach calls a "determinate system with combinatorial behaviour" (Klir & Valach 1967).

It seems increasingly clear that it is well worth following a path of research into the abstract structures that account for, and extend, knowledge. For the social systems the goal looms in the form of a deterministic system of formulae. Maybe today this is difficult to attain completely, but the endeavours towards this goal give valuable partial results, and give a clear insight into essentially new areas. We may conclude this section on general systems theory by reminding the reader of Laplace's famous pronouncement that, in principle, it ought to be possible to predict peoples' behaviour with the help of systems of differential equations. This gives some idea about the confidence felt in, and the value of, abstraction.[1]

4.2.6. *Simulation with the model systems*

The penetration into the model system for the assessment of its characteristics can be made analytically, numerically or through a combination of these two ways. By numerically is here understood calculations and handling in a non-symbolical way, for example, variables are prescribed or are given numerical values. For simple systems, whose implications are easily investigated, it is sufficient to proceed by mathematical and statistical deductions. An example of such a parsimonious form is the logistic system that has been dealt with in section 4.2.1.

The present study, however, is mainly concerned with numerical analysis, depending on the complexity of the model studied. In general, system S has several objects with interactive traits, apart from which many variables are included in the mechanisms. This means that numerical methods, besides the analytical methods, must be applied to the model. This operation is called simulation, and, as has already been mentioned in Chapter 2, it means the observation of the model's conduct during experimentation.

The stochastic nature of system S implies that by simulation we are in the present case confronted with the problem of sampling from probability distributions. For this we use random numbers and when simulating with the model system S we include elements from which random numbers are generated. The situation is depicted in Fig. 4.3.

The use of random numbers is generally referred to as the Monte Carlo method. This is a wide subject for discussion and has been given considerable treatment in the literature. We refer to Tocher (1963) and Hammers-

---

[1] This well-known pronouncement by Laplace (1812) is referred to here to emphasize the challenge it makes to model building and forecasting, and not so much to emphasize deterministic aspects.

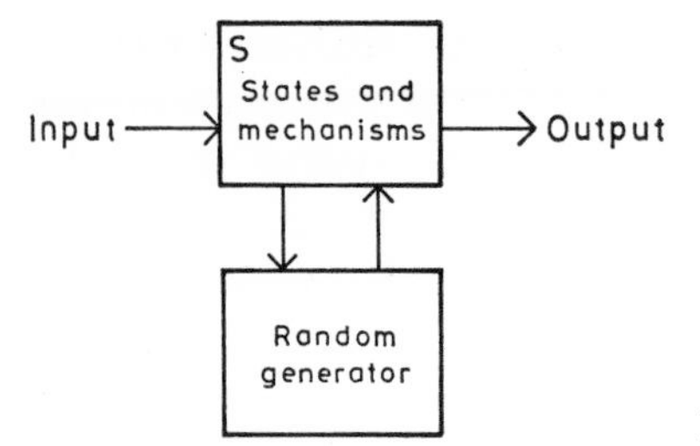

Fig. 4.3. Simulation with system S with the help of a random number generator.

ley & Handscomb (1964) for descriptions of the use of various Monte Carlo techniques and to Naylor (1971) for a review of the method in the context of computer simulation models.

The use of Monte Carlo simulation in the present case may be illustrated as follows. Suppose that a mechanism formulated as a conditional statistical distribution $f(\eta, \xi | \zeta)$ of the state of, and the output from, a system given an input ζ is aimed at. Assume that a simple description of this distribution is not available, but, however, the conditional marginal distributions $g(\eta | \xi)$ and $h(\xi | \eta, \zeta)$ are given in the form of analytical expressions. From the following fundamental rule concerning conditional expectations

$$f(\eta, \xi | \zeta) = g(\eta | \zeta) h(\xi | \eta, \zeta) \tag{4.2.42}$$

observations from the distribution $f(\eta, \xi | \zeta)$ can be generated with each observation obtained in two steps: first an observation on the state variable is obtained through a random drawing from $g(\eta | \zeta)$, then an observation on the output variable is obtained through a random drawing from $h(\xi | \eta, \zeta)$. Thus, in the absence of an expression for $f(\eta, \xi | \zeta)$, sampling distributions for it can be obtained by the Monte Carlo method.[1]

A special problem that occurs in connection with the use of this method is that the generators used for the production of intended sequences of random numbers exhibit non-random properties. This problem, however, will not be the subject of any closer study in this report. In the applications the generators that the computers in question are provided with will therefore be viewed as producing variable values that are random. For the study of occurring random number generators, reference is made to Naylor (1971) who gives, among other things, references to texts on the underlying number theory.

4.3. The modelled systems

In this section we focus our interest on members in the class of social systems that, as regards behavioural aspects, may be described in terms of the model system S. The modelled system, which is thus assumed to be

[1] This argument leans heavily on Orcutt et al. (1961), where a similar use of a factoring of a probability distribution is demonstrated.

a real system, is represented by S^R. It is assumed that different objects of this system can be identified through classification schemes. It is also assumed that variables and variable values in the real system are definable through, and obtainable from, measuring rules for defined objects in the system. The structure of the real and modelled system, however, is not assumed to be explicitly observable. One of the incentives for modelling in this work is precisely this condition.

All we can do, therefore, is to study the behaviour of the modelled system on the basis of hypotheses as to how the behaviour is generated. As will become clear in the next section, the behaviour is assumed to follow the same rules as the behaviour of the model system. That is, the observations are interpreted as observations on random variables generated by the probability distributions found for the structure of system S.[1]

Just as observations on the model system's behaviour are obtained through simulation experiments, the observations on the modelled system's behaviour may thus be viewed as the outcome of some fictitious experiments.[2]

4.4. Specification of the empirical part of models

The basis for the specification of the empirical part of a simulation model (section 2.3) is naturally founded on anticipated similarities between the model and the modelled system. In this respect cybernetics states the importance of models which build on behavioural similarities: "... it is the model of behaviour which is particularly typical of cybernetics, and we may rightly say that this approach to modelling is one of the chief contributions of cybernetics to the general methodology of science." (Klir & Valach 1967, p. 93.)

With regard to the behavioural aspects of the two systems, only variables which we interpret as state variables are dealt with in the theoretical part

---

[1] To link up with the distinctions made in section 2.3, the variables assumed to be operationally defined above are the evidential interpretations, whereas the corresponding and underlying stochastic variables are the referential interpretations. In the case of completely valid measuring instruments we should have a complete match between "reference" and "evidence" in the sense that obtained observations can be regarded as observations on the stochastic variables. So that we shall not have to assume this we shall introduce a "measurement error" (in general unobservable) reflecting a difference between "reference" and "evidence" in the treatment of observations as, for example, in relation (5.2.22) in the next chapter. The treatment of "measurement errors" is ventilated to some extent in section 7.4.

[2] Wold (1966) launches "fictitious experiments" as a means for the interpretation of observational (non-experimental) data in causal terms. If his view is adopted here we may thus attribute to the model values relating to causal explanations of the empirical phenomena studied. This, of course, in those cases where the modelling attempts are deemed "successful".

of this report. The limitations thus imposed on the report are, to some extent, compensated in those cases where it is shown to be possible to enlarge the state variables with variables which give an adequate picture of what we interpret as input and output variables.

As for the model system, the notation system closely follows that used in section 4.2. To indicate that the structure is now used in a model context we put the index "M" on the variables. A vector of state variables of the objects in system S at time-point τ is represented by $\boldsymbol{\eta}_{M\tau}$. The vector of all inputs from the environment S_0 during the observed time-period is written $\boldsymbol{\zeta}_M$.

Assume that the corresponding observable variables in the modelled system are defined. Designate them with $\boldsymbol{\eta}_{Rt}$, where t designates some real time, and $\boldsymbol{\zeta}_R$. As described in the previous section, the variable $\boldsymbol{\eta}_{Rt}$ is regarded as a stochastic variable.

Consequently, similarities in behaviour are naturally founded on relations between the variable $\boldsymbol{\eta}_{M\tau}$ in the model system and the variable $\boldsymbol{\eta}_{Rt}$ in the modelled system for comparable circumstances. If the stochastic nature of these two variables is taken into account, the following specification of a behavioural model is thus a straightforward interpretation; cf. Klir & Valach (1967).

Definition 4.6. *System S is a valid behavioural model of system S^R if there exist functions $\mathbf{f}_\eta$, $\mathbf{f}_\zeta$ and f_τ such that:*

- *for every pair of initial states $\boldsymbol{\eta}_{M0}$ and $\boldsymbol{\eta}_{R0}$ of the two systems that fulfils the relation*

$$\boldsymbol{\eta}_{M0} = \mathbf{f}_\eta(\boldsymbol{\eta}_{R0}), \tag{4.4.1}$$

- *for every pair of inputs $\boldsymbol{\zeta}_M$ and $\boldsymbol{\zeta}_R$ to the two systems that fulfils the relation*

$$\boldsymbol{\zeta}_M = \mathbf{f}_\zeta(\boldsymbol{\zeta}_R), \tag{4.4.2}$$

- *for each set of corresponding time-points τ_i and t_i in the two systems that fulfils the relation*

$$\tau_i = f_\tau(t_i), \quad i = 1, ..., T, \tag{4.4.3}$$

where the time-scales for τ and t are assumed to be placed on the positive real axis with starting points at the origin,

- *the probability distributions of the two stochastic variables*

$$\boldsymbol{\eta}_M = \begin{pmatrix} \boldsymbol{\eta}_{M\tau_1} \\ \vdots \\ \boldsymbol{\eta}_{M\tau_T} \end{pmatrix} \quad \text{and} \quad \boldsymbol{\eta}_R = \begin{pmatrix} \mathbf{f}_\eta(\boldsymbol{\eta}_{Rt_1}) \\ \vdots \\ \mathbf{f}_\eta(\boldsymbol{\eta}_{Rt_T}) \end{pmatrix} \tag{4.4.4}$$

are the same.

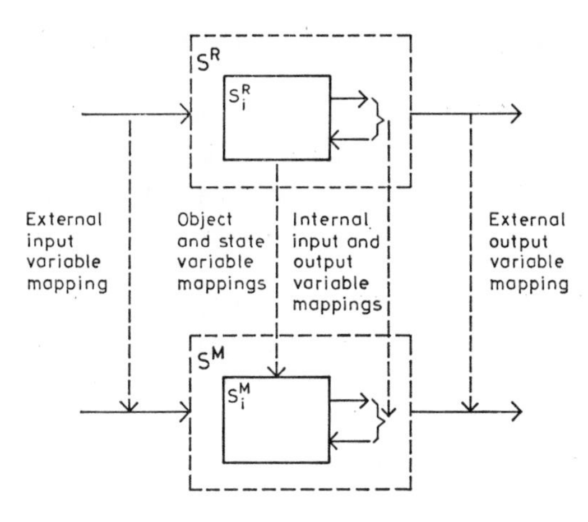

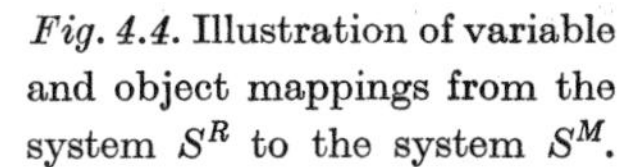

Fig. 4.4. Illustration of variable and object mappings from the system S^R to the system S^M.

The system S, provided with this specification—or if there exist intentions, expectations or articulated propositions regarding the specification—is represented by S^M and will for convenience be referred to as the (behaviour) model.[1]

If system S^M is a behavioural model of S^R it follows that the mean-values, variances etc. of the state variables in the model are equal to corresponding characteristics of the state variables transformed by $\mathbf{f}_\eta$ in the real system. Inferences about the modelled system can thus be made with the help of the model.

In other words, the structure of the model system can be used as a substitute and prototype for the structure of the modelled system, which is assumed to be unknown or not directly observable. And instead of carrying out experiments or survey operations in the modelled system we may obtain the same results through simulation with the model. It is in the light of these perspectives that the problem of testing the model system based on Definition 4.6 is given considerable attention in the subsequent chapter.

The above specification is adequate for the purposes of this study, though, of course, only one among many conceivable specifications. Other and less far-reaching model specifications, on the one hand, are more appropriate for use in particular contexts. For example, when concentrating on a "mean-value" model one obtains: "... the mean of the variable $\boldsymbol{\eta}_M$ is equal to the mean of the variable $\boldsymbol{\eta}_R$" etc. Such limitations do not, however, alter the main lines of the argument in what follows.

For more penetrating model concepts, on the other hand, one can define object mappings and, for every resulting pair of objects, define "variable mappings" corresponding to the functions $\mathbf{f}_\eta$ and $\mathbf{f}_\zeta$. In Fig. 4.4 a

[1] We have thus chosen a "static" formulation of the model specification in order to simplify the exposition. A more "dynamic" formulation would use the theory of stochastic processes.

conceivable and many-faceted mapping between the system S^M and the system S^R is suggested.

The handling of these more ambitious models is, of course, a problem far more difficult to penetrate and deal with. Reference is here made to the model specifications employed in cybernetics concerning morphological similarities between model and modelled systems as described e.g. in Klir & Valach (1967).

In conclusion, we make the observation that only variable and object aspects are used in the model specification above. This makes many epistemological problems simpler, since structural aspects such as mechanisms and parameters, which were used in the definition of the model structure, and which in general are not observable, need not be interpreted in relation to the modelled system.[1]

---

[1] As regards the specification of the theoretical part of the model (section 2.3), however, the mechanisms and parameters in the structure S are interpreted on the basis of an underlying theory. While stress is laid on the model system's behaviour in specifying the model's empirical part, the structural aspects of the model system thus belong more to its theoretical part; cf. Bråten & Norlén (1969). From cybernetics we know that in general there is a set of model structures having the same behaviour (Ashby 1952). The selection of one particular model structure among possible behaviour-equivalent structures is therefore made, in the light of this discussion, on theoretical grounds. This reasoning applies in econometrics. When concentrating on the empirical part, on the one hand, the behaviour as given by the reduced form occupies the focus of interest. On the other hand, when one considers the theoretical part, the structural form is focussed. Reference is made to section 5.3, Chapter 9 and to the Appendix for a further discussion of this subject.

Chapter 5

Testing of simulation models

5.1. Introduction

The primary concern of this chapter is to describe testing procedures of simulation models. Testing a simulation model will, in the cases that will be adduced, refer to judgments of the model, as well as investigations on the extent to which the proposed conditions with reference to the embedded criteria in a model specification are fulfilled. The term validation will be used as a synonym for the notion testing. To test or validate a model is therefore, in general terms, to estimate how well the model matches its specification.

In the case of simulation models, the testing problem is dealt with in the literature as more or less dependent upon the problem of model specification (Forrester 1961). Different opinions are put forward, but in many cases the differences may be derived from the absence of a distinct concept of model.

As has been made clear in section 2.3, a model is, as a matter of course, a construct and different from what is modelled. As a result of this, as is to be expected, differences arise between the model system and the modelled system, i.e. a model never completely reproduces the image of the modelled system; cf. Bunge (1967: I). A well-formulated model, with the intended representational qualities clearly defined, constitutes, however, an ideal starting point for a penetrating study of the problem of validation.

Validity is interpreted as the close agreement between the model and its specification. Following on from this it should be possible, in the cases where negative results are obtained, to revise the model by altering its specification when stating the model's properties, besides the given possibility of altering the model's stucture; cf. Wold (1969*a*).

In accordance with what has been said in previous chapters, we shall mainly be concerned with testing the empirical part of simulation models. The material in this chapter pertaining to this problem is given in the next section.

In section 5.3 the theoretical aspects in connection with validation are briefly discussed. Finally, in section 5.4, the problem of validating general models is introduced.

5.2. Testing the empirical part of models

5.2.1. *Points of departure*

The testing of the empirical part of a model can be symbolically illustrated in the following three steps. First one formulates a (composite) hypothesis based on, and developed from, the model specification and arranging one or more distribution qualities. Thereafter two sets of values $\mathbf{y}_M$ and $\mathbf{y}_R$ are obtained

$$\mathbf{y}_M = (O_M, S^M) \tag{5.2.1}$$

$$\mathbf{y}_R = (O_R, S^R), \tag{5.2.2}$$

which are either primary and original observations or compiled, maybe even in some respects tested, characteristics from the operations O_M and O_R on the model system S^M and the modelled system S^R respectively. Then a test of fit, using the following function

$$V(\mathbf{y}_M, \mathbf{y}_R \,|\, O_M, O_R, S^M, S^R), \tag{5.2.3}$$

is connected to the obtained variables and the given conditions. This function may be regarded as a Boolean function with two values: either a "true" value, meaning that $\mathbf{y}_M$ is reasonable in relation to $\mathbf{y}_R$, or a "false" value, meaning that $\mathbf{y}_M$ is not reasonable in relation to $\mathbf{y}_R$. The discussion of model testing will, in what follows, be carried on in terms of the two kinds of operation O_M and O_R and the function V.

Following the principles of significance and goodness-of-fit testing, the function (5.2.3) incorporates an evaluation procedure to ascertain whether the values $\mathbf{y}_M$ and $\mathbf{y}_R$ differ from each other by an amount greater than that attributable to random variation alone.[1]

For this purpose the distribution of the statistic corresponding to, for example, the difference between $\mathbf{y}_M$ and $\mathbf{y}_R$ must first be calculated on the basis of the hypothesis that the model system is a behaviour model of the real system in question (the null hypothesis). In line with the rules of hypothesis testing, a small probability p is then chosen before the observations $\mathbf{y}_M$ and $\mathbf{y}_R$ are made. This probability p is accordingly put into the function V as one of its arguments. Finally, if the value of the difference between $\mathbf{y}_M$ and $\mathbf{y}_R$ has a smaller probability than p of being attained or exceeded, the function V attains the value "false" and the null hypothesis is rejected at the 100 p% significance level; otherwise the function V attains the value "true" and the null hypothesis is retained, at least for the time being.

If these rules are applied, the error of Type I, i.e. the error of asserting a departure from the null hypothesis when it is actually true, and the error of

[1] For an authoritative exposition of the theory of statistical testing, see Cramér (1945) or Kendall & Stuart (1961).

Type II, i.e. the error of not asserting a departure from the null hypothesis when it is actually false come into the picture. For obvious reasons it is desirable to construct a test where the probabilities of comitting these two kinds of error are small. As for the error of Type I, the probability of making this error is under control, since it is equal to p. Instead, the problem concerns the error of Type II. In the present case this problem is complicated, owing to the fact that the alternative hypothesis may hardly be taken into operational account during the test construction. Some attention may, however, be given to the construction of a test which is sensitive to departures from the null hypothesis, i.e. a test which has a large power, since the power of a test is defined as the complement of the probability of making the error of Type II.

5.2.2. *Turing tests*

It should at this juncture be emphasized that it is often costly and difficult to set up, and explicitly define, functions V that have a high power, and with which relevant operations O_M and O_R interplay. One possibility is to avoid explicit formulation. Then this should be replaced by information about the two systems intuitively judged as important, and by intuitive judgement. In the respects concerned, such a procedure will be expressive for a model that is not completely developed, as it allows the possibility of finding out existing defects of the model relatively quickly.

In this connection it is of special interest to give an account of an idea by Turing (1950). Originally his idea, in the form of the so-called Turing tests, was advanced in connection with the evaluation of artificial intelligence models. Briefly, a Turing test may be described as follows: a chosen person is confronted with both the model and what has been modelled. The person is asked to put questions concerning the two systems, and in this way to try and get a clear idea as to which system is the model. This idea is here difficult to put into practice. It is, however, possible to choose another additional variant, which to some extent fulfils Turing's intentions, as is presented below.

A chosen person may, in a prepared situation, be confronted with pre-arranged alternatives. In each situation two kinds of information are given: observations from the model and observations from the modelled. The task of the respondent is to decide on an alternative, and to state which observations belong to the modelled system according to his understanding. The purpose of the experiment is thus to find out to what extent it is possible to discriminate between the model and the modelled. Naturally, the resulting values from this experiment are closely related to the design and proportions to which the given information is relevant, as well as to the way in which the initial information was accounted for, all in relation to the goal of the work with the model. The result of the experiment is thereafter tested with the usual statistical methods. If, for example, the person's answers consist of n paired comparisons, it may be reasonable to arrange the following one-sided test, with the null hypothesis

$$H_0 : P = \tfrac{1}{2} \tag{5.2.4}$$

and the alternative hypothesis

$$H_1 : P > \tfrac{1}{2}, \tag{5.2.5}$$

where P is the probability of making the right decision between the alternatives at each choice. In other words the null hypothesis states that the person chooses completely at random. If the person chooses f correct alternatives, the null hypothesis is rejected when $f \geqslant f_0$, where f_0 is the smallest integer for which the inequality

$$\sum_{i=f_0}^{n} \binom{n}{i} \left(\frac{1}{2}\right)^i \left(\frac{1}{2}\right)^{n-i} > p \tag{5.2.6}$$

holds, where p is the significance level. If the null hypothesis is rejected, this means that there exists a reason to question the model's correct representation of the modelled in the respects considered.[1]

Several extensions and digressions of this method are, of course, possible. In the general case choice situations are not of similar kinds. Among other things, it is a realistic possibility that different kinds of information are supplied, and that this might even refer to different conditions. Furthermore, the experiment may, at the same time, be enlarged to include several persons. Moreover, there are obvious reasons for assuming that the persons in question ought to be split into groups with reference to their different qualifications and backgrounds. An interesting and rewarding situation, with regards to the distribution of results, may arise if the persons in the different groups act as followers of different authorities or schools of thought. In such circumstances it is reasonable to describe the situation in terms of the following theoretical model

$$P_{ij} = \mu + \alpha_i + \beta_j, \tag{5.2.7}$$

where P_{ij} represents the probability of making a correct choice in situation number i, for a person belonging to the jth group. For further treatment of the analytic model (5.2.7) we refer to the analysis of variance.[2] It is also possible to include, amongst other things, effects that reflect the interactions between groups and choice situations. Finally, this intuitive evaluation has of course variants of differing value, because different forms of wrong ideas may occur, but cannot always be discovered.

[1] This framing of the experimental situation is comparable with the problem treated in part II in Fisher (1966 (1935)).

[2] For the groundwork of the analysis of variance, see Scheffé (1959). A simplified exposition of the subject is given by Brownlee (1960).

5.2.3. *Goodness-of-fit tests*

In a more systematic study of (5.2.1–3) with the model specification given by Definition 4.6 as starting point, the task of testing for similarities between the distributions of the two variables in (4.4.4) is discerned. We shall describe how a chi-square test may be applied to ascertain whether the two mean vectors are in statistical agreement with each other.

The problem is first reduced by the assumption that the variables of the modelled system are transformed according to the functions $\mathbf{f}_\eta$, $\mathbf{f}_\zeta$ and f_τ in (4.4.1–4). For the sake of simplicity the old notation for the modelled system's variables is retained. For the rest, the notation system follows closely that employed in sections 4.2–4.

In the following example, a general illustration is given of the type of test immediately in view, which will be dealt with here.

Example 5.1. For the purpose of illustration it is assumed that the true values for parameters and predetermined variables of the model are known, and also that all the observations, from both the real system and the model, are free from any measurement errors. Denote the set of state variables in the two systems, at a number of observed time-points, with $\boldsymbol{\eta}_M$ and $\boldsymbol{\eta}_R$ respectively.

A testing implication that follows from the model specification is the testing of whether the mean values for the two distributions of $\boldsymbol{\eta}_M$ and $\boldsymbol{\eta}_R$ are equal.

Assume that $N_M(N_R)$ mutually independent measurements of $\boldsymbol{\eta}_M(\boldsymbol{\eta}_R)$ are obtained from the model (modelled system). A statistical method for testing the equality of mean values is, assuming a normal distribution, Hotelling's

$$T^2 = \frac{N_R N_M}{N_R + N_M} (\bar{\mathbf{y}}_M - \bar{\mathbf{y}}_R)' \mathbf{S}^{-1} (\bar{\mathbf{y}}_M - \bar{\mathbf{y}}_R), \tag{5.2.8}$$

where $\bar{\mathbf{y}}_M$ and $\bar{\mathbf{y}}_R$ are the observed mean values, and where

$$\mathbf{S} = \frac{1}{N_M + N_R - 2} \left(\sum_{i=1}^{N_M} (\mathbf{y}_{Mi} - \bar{\mathbf{y}}_M)(\mathbf{y}_{Mi} - \bar{\mathbf{y}}_M)' + \sum_{i=1}^{N_R} (\mathbf{y}_{Ri} - \bar{\mathbf{y}}_R)(\mathbf{y}_{Ri} - \bar{\mathbf{y}}_R)' \right). \tag{5.2.9}$$

It can be shown that under the null hypothesis the quantity

$$F = \frac{N_R + N_M - n - 1}{(N_R + N_M - 2)n} T^2 \tag{5.2.10}$$

has the variance ratio F distribution with degrees of freedom n and $N_R + N_M - n - 1$, where n is the number of elements in the state vector. This test concerning equal population mean vectors is described, for example, in Morrison (1967).

It is thus possible, under the given conditions, to perform a relatively simple test.

The model's parameters and predetermined variables are, however, estimated quantities. Furthermore it is plausible to assume that the observed values from the real system are affected by measurement errors.[1] In several cases, therefore, the assumptions used in the above example are unrealistic. We need results from analyses of the model, as well as of the modelled system, before ambitious attempts at validation can be tried.

Investigation of the necessary analyses is undertaken in connection with the relation representation (4.2.37–38) of the model system's behaviour. The aim is in the first place to obtain a statistic corresponding to (5.2.10) for testing the equality of mean values from more realistic assumptions.

The estimate of the mechanism parameter $\boldsymbol{\alpha}$ is called $\mathbf{a}$. The estimate of the initial value $\boldsymbol{\eta}_0$ for the studied state variables at time-point zero is designated by $\mathbf{y}_0$. Furthermore, we designate by $\mathbf{z}$ the estimate of the vector $\boldsymbol{\zeta}$ of all the inputs from the neighbourhood during the observed time-period.

In conformity with the notation in Example 5.1, the chosen state variables in the two systems are given at a number of time-points as $\boldsymbol{\eta}_M$ and $\boldsymbol{\eta}_R$. We assume that the operations (5.2.1–2) consist of

O_M: Mutually independent observations without measurement errors from operations O_R: 1–3 (see below) estimated model

$$\mathbf{y}_{Mi} \quad i = 1, \ldots, N_M. \tag{5.2.11}$$

O_R: Measurements and observations which give

1. estimate $\mathbf{a}$ of $\boldsymbol{\alpha}$
2. estimate $\mathbf{y}_0$ of $\boldsymbol{\eta}_0$
3. estimate $\mathbf{z}$ of $\boldsymbol{\zeta}$
4. mutually independent observations

$$\mathbf{y}_{Ri} \quad i = 1, \ldots, N_R. \tag{5.2.12}$$

From section 4.2.4 it follows that for the state variables in the estimated model there is a relation

$$\boldsymbol{\eta}_M = \boldsymbol{\phi}(\mathbf{a}, \mathbf{y}_0, \mathbf{z}) + \boldsymbol{\epsilon}_M(\mathbf{a}, \mathbf{y}_0, \mathbf{z}), \tag{5.2.13}$$

where

$$E(\boldsymbol{\eta}_M | \mathbf{a}, \mathbf{y}_0, \mathbf{z}) = \boldsymbol{\phi}(\mathbf{a}, \mathbf{y}_0, \mathbf{z}). \tag{5.2.14}$$

For the true values of the parameters and predetermined variables the corresponding relations are

[1] An exposition of the theory of parameter estimation adapted to the present problem is given in section 7.3. The general problem of analysing survey errors, including measurement errors, is ventilated in section 7.4.

$$\tilde{\boldsymbol{\eta}}_M = \boldsymbol{\phi}(\boldsymbol{\alpha}, \boldsymbol{\eta}_0, \boldsymbol{\zeta}) + \tilde{\boldsymbol{\epsilon}}_M(\boldsymbol{\alpha}, \boldsymbol{\eta}_0, \boldsymbol{\zeta}), \tag{5.2.15}$$

where

$$E(\tilde{\boldsymbol{\eta}}_M | \boldsymbol{\alpha}, \boldsymbol{\eta}_0, \boldsymbol{\zeta}) = \boldsymbol{\phi}(\boldsymbol{\alpha}, \boldsymbol{\eta}_0, \boldsymbol{\zeta}). \tag{5.2.16}$$

The sample means for the state variables in the model system are, using (5.2.13) written

$$\bar{\boldsymbol{\eta}}_M = \boldsymbol{\phi}(\mathbf{a}, \mathbf{y}_0, \mathbf{z}) + \frac{1}{N_M} \sum_{i=1}^{N_M} \boldsymbol{\epsilon}_{Mi}, \tag{5.2.17}$$

where $\boldsymbol{\epsilon}_{Mi}$ is the disturbance for the ith observation.[1]

The first term on the right-hand side depends upon the estimated quantities of the model system. Using (5.2.16) we can write (5.2.17) as follows

$$\bar{\boldsymbol{\eta}}_M = \boldsymbol{\phi}(\boldsymbol{\alpha}, \boldsymbol{\eta}_0, \boldsymbol{\zeta}) + \boldsymbol{\epsilon}_\phi + \frac{1}{N_M} \sum_{i=1}^{N_M} \boldsymbol{\epsilon}_{Mi}, \tag{5.2.18}$$

where

$$\boldsymbol{\epsilon}_\phi = \boldsymbol{\phi}(\mathbf{a}, \mathbf{y}_0, \mathbf{z}) - \boldsymbol{\phi}(\boldsymbol{\alpha}, \boldsymbol{\eta}_0, \boldsymbol{\zeta}). \tag{5.2.19}$$

A mean value obtained from the model system may thus be understood as consisting of three terms. The first term on the right-hand side of (5.2.18) gives the conditional expectation of the state from the hypothetical model system, with the true parameter values. The second term is due to the fact that the model is estimated. This term originates from the conditions obtaining for the operations O_R: 1–3. We designate this term the propagated error; cf. Mandel (1964). Finally, the third term is dependent on the stochastic elements in the estimated model. This term may thus be said to originate from the operation O_M. On the basis of the results hitherto obtained and described, it is clear that the analyses of the model system are determined by the needs of estimating the distribution characteristics for the propagated error $\boldsymbol{\epsilon}_\phi$ and the disturbance $\boldsymbol{\epsilon}_M$.

The requirements for the analyses of the modelled system are described as starting from the assumed conditional distribution of $\boldsymbol{\eta}_R$ for given $\boldsymbol{\alpha}$, $\boldsymbol{\eta}_0$ and $\boldsymbol{\zeta}$; cf. section 4.3. The following decomposition of $\boldsymbol{\eta}_R$ can be made

$$\boldsymbol{\eta}_R = \boldsymbol{\phi}(\boldsymbol{\alpha}, \boldsymbol{\eta}_0, \boldsymbol{\zeta}) + \boldsymbol{\epsilon}_R(\boldsymbol{\alpha}, \boldsymbol{\eta}_0, \boldsymbol{\zeta}_0), \tag{5.2.20}$$

where

$$E(\boldsymbol{\eta}_R | \boldsymbol{\alpha}, \boldsymbol{\eta}_0, \boldsymbol{\zeta}) = \boldsymbol{\phi}(\boldsymbol{\alpha}, \boldsymbol{\eta}_0, \boldsymbol{\zeta}) \tag{5.2.21}$$

---

[1] For the sake of simplicity the arguments given in parenthesis in $\boldsymbol{\epsilon}_M$ $(\mathbf{a}, \mathbf{y}_0, \mathbf{z})$, indicating that the distribution of the disturbance might depend on them, are omitted in relation (5.2.17). To the extent deemed suitable the same rule applies for the representation of other disturbance terms in what follows.

defining the disturbance term $\boldsymbol{\epsilon}_R$. The variable $\boldsymbol{\eta}_R$ is thus assumed to have genuine variability corresponding to the genuine variability of the variable $\boldsymbol{\epsilon}_M$ in the model system.

An observation $\mathbf{y}_{Ri}$ on $\boldsymbol{\eta}_R$ can be written

$$\mathbf{y}_{Ri} = \tilde{\mathbf{y}}_{Ri} + \mathbf{v}_{Ri}, \tag{5.2.22}$$

where $\tilde{\mathbf{y}}_{Ri}$ is an (unobserved) value for $\boldsymbol{\eta}_R$ and where $\mathbf{v}_{Ri}$ is a value for $\boldsymbol{\upsilon}_{Ri}$, the measurement error in the ith observation.[1]

It follows that the mean value of the measurements (5.2.12) is distributed as

$$\bar{\boldsymbol{\eta}}_R = \boldsymbol{\phi}(\boldsymbol{\alpha}, \boldsymbol{\eta}_0, \boldsymbol{\zeta}) + \frac{1}{N_R}\sum_{i=1}^{N_R} \boldsymbol{\epsilon}_{Ri} + \frac{1}{N_R}\sum_{i=1}^{N_R} \boldsymbol{\upsilon}_{Ri}, \tag{5.2.23}$$

where $\boldsymbol{\epsilon}_{Ri}$ is the disturbance in the ith observation, reflecting the genuine variability.

Thus, in the first place, the analyses in the modelled system need to be directed towards the study of the distributions, partly for the genuine variability in the system, partly for the measurement errors in the observations from the system, which are reflected by $\boldsymbol{\epsilon}_R$ and $\boldsymbol{\upsilon}_R$ respectively in the above relation.

When the results are summarized, a plausible aim for the analyses consists of an estimation of the mean values and dispersion matrices of the four variables $\boldsymbol{\epsilon}_\phi$, $\boldsymbol{\epsilon}_M$, $\boldsymbol{\epsilon}_R$ and $\boldsymbol{\upsilon}_R$, which are statistically independent under quite general conditions.

Example 5.2. The following simple model mechanism, in the form of a conditional frequency function is used as an illustration of the necessary analyses

$$f(\tilde{y} \mid \alpha, \zeta) = \frac{1}{\sqrt{2\pi}} \exp\{-\tfrac{1}{2}(\tilde{y} - \alpha\zeta)^2\}, \tag{5.2.24}$$

with scalar quantities $\tilde{y}$, ζ and α. The estimated frequency function is written as

$$f(y \mid a, z) = \frac{1}{\sqrt{2\pi}} \exp\{-\tfrac{1}{2}(y - az)^2\}. \tag{5.2.25}$$

Assume, for the sake of simplicity, that all measurements of both the outcome from the model and the real system are performed without measurement errors.

This simple situation immediately reveals that the propagated error (5.2.19) can by using (5.2.13–16) be written as

$$\varepsilon_\phi = (a - \alpha)z. \tag{5.2.26}$$

[1] For an interpretation of measurement error; see section 4.3.

If the parameter α is estimated by the method of least squares, that is if

$$a = \sum y_i z_i / \sum z_i^2, \tag{5.2.27}$$

one obtains

$$E(\varepsilon_\phi) = 0 \quad D(\varepsilon_\phi) = z^2 / \sum z_i^2. \tag{5.2.28}$$

If one considers the mechanism (5.2.25), one finds that the disturbance ε_M has a mean value of zero, and a variance of one. If the mechanism is correct, it is further seen that the disturbance ε_R also has mean zero and unit variance. The assumptions yield the final result that the measurement error v_R is equal to zero.

The above conclusions could be obtained directly from consideration of the model mechanism. It may be noted that, for reasons which have been stated in section 4.2, this possibility of analytically deducing characteristics of the model behaviour does not usually exist. This comment is also true for Example 5.3 below.

Assuming that the analysis can be carried out successfully, a partial goal in the form of an all-encompassing test of the validity of the model is discernible. On the assumption that the distribution of the behaviour of the model and the real system are the same, a need arises to test the null hypothesis that the mean values in the model and modelled systems are the same. When using (5.2.18) and (5.2.23) the null hypothesis implies

$$\bar{\boldsymbol{\eta}}_M - \bar{\boldsymbol{\eta}}_R = \boldsymbol{\epsilon}, \tag{5.2.29}$$

where

$$\boldsymbol{\epsilon} = \boldsymbol{\epsilon}_\phi + \frac{1}{N_M} \sum_{i=1}^{N_M} \boldsymbol{\epsilon}_{Mi} - \frac{1}{N_R} \sum_{i=1}^{N_R} \boldsymbol{\epsilon}_{Ri} - \frac{1}{N_R} \sum_{i=1}^{N_R} \boldsymbol{\upsilon}_{Ri}. \tag{5.2.30}$$

With reference to the test and validation function (5.2.3) the following logical expression can be advanced, assuming a normal distribution,

$$V = \{(\bar{\mathbf{y}}_M - \bar{\mathbf{y}}_R)' D(\boldsymbol{\epsilon})^{-1} (\bar{\mathbf{y}}_M - \bar{\mathbf{y}}_R) \leqslant c_p\}, \tag{5.2.31}$$

where c_p is the critical value at significance level p, which is obtained from the non-central Chi-squared distribution with non-centrality parameter $E(\boldsymbol{\epsilon})' E(\boldsymbol{\epsilon})$ and with n degrees of freedom, where n is the number of elements in the state vectors.

By making reasonable demands upon the statistical accuracy, it can be seen that the test function (5.2.31) should be usable from, to a certain extent controlled, results of analyses in the form of estimates of the mean values and the dispersion matrices for the four variables on the right-hand side of (5.2.30).

Example 5.3. If reference is made to the simple model mechanism in Example 5.2, an illustration of the test (5.2.31) is obtained. If the results in the example are summarized the following is obtained for the difference of the mean values (5.2.29)

$$E(\varepsilon)=0 \quad D(\varepsilon)=\frac{1}{N_M}+\frac{1}{N_R}+\frac{z^2}{\Sigma z_i^2} \tag{5.2.32}$$

if the mechanism (5.2.24) is correct.

In order to illustrate the use of this test function two Monte Carlo studies have been carried out with the following experimental design.

- $p=0.05$
- $N_M=N_R=10$
- The estimate (5.2.27) was made from 11 observations with $z_i=i-6$; $i=1, ..., 11$
- $c_p=\chi^2_{0.05}(1)=3.84$
- The test was performed 100 times for each of the following 11 values of z

  $z = -5, -4, ..., 5,$

  i.e. in all 1 100 tests.

In the first study the mechanism was correct. The proportion of rejections, that is the proportion of times the function V attained the value "false", for the 11 values of z, varies between 0.02 and 0.08, with a mean value of 0.05.[1] This result reveals that no demonstrable difference from that which was expected arises. The average number of rejections is seen to correspond to the significance level.

In the second study the following mechanism was used

$$f(\tilde{y}\,|\,\alpha,\zeta)=\frac{1}{\sqrt{2\pi}}\exp\{-\tfrac{1}{2}(\tilde{y}-1-\alpha\zeta)^2\}, \tag{5.2.33}$$

but was incorrectly specified as (5.2.24). The information obtained from the Monte Carlo study reveals that the proportion of rejections, which in this case is the correct decision, varies between 0.35 and 0.84, with a mean value of 0.61.

A testing implication from the model specification has thus been treated which suggests an investigation to find out to what extent it is probable that the mean values of the behaviour in the two systems are equal.

The method described also appears to be applicable in the continued testing for other similarities in the distributions of the variables in the two systems, according to the model specification.

---

[1] The computations in the Monte Carlo study were performed on the university computer CDC 3600 in Uppsala.

At this juncture the desirability of studying the genuine variability of the two systems and attempting to compare the dispersion matrices of the disturbance $\tilde{\boldsymbol{\epsilon}}_M$ in (5.2.15) and $\boldsymbol{\epsilon}_R$ in (5.2.20) becomes obvious. This problem and other similar problems are not, however, dealt with in this work.

At the end of section 4.4 the idea was put forward of enlarging the model concept to include object mappings between the systems.

It is possible to discuss, in terms of the operations (5.2.1–3), tests for the proposed object mappings. Thus an object mapping from system S^R to system S^M exists if the operations O_M and O_R can be described in such a way that one obtains classifications and statements of correspondences between objects in the two systems.

If, for example, each object of the two systems can be given an unequivocal equivalent in a natural number in the vectors $\mathbf{y}_M$ and $\mathbf{y}_R$ respectively, and if the objects which have been allotted the same natural numbers are seen to constitute each other's equivalents, then there exists an object mapping from the system S^R to the system S^M if the natural numbers in the vector $\mathbf{y}_M$ are also found in the vector $\mathbf{y}_R$. The two vectors $\mathbf{y}_R$ and $\mathbf{y}_M$, together with the rules ascribed to them, the operations, also constitute a description of the object mapping.

5.2.4. *A review of testing approaches*

A distinctive feature of the current discussion on the validation of simulation models is that the problem often resolves itself into attitudes to the model concept; cf. the surveys of the positions in the literature by Larsson & Lundin (1970*a–b*). One can observe a hierarchy of different meanings given to the model concept in the literature, which it falls outside the scope of this monograph to penetrate. We confine ourselves to the following general comments on others' views on validation.

It is first pointed out that there does not seem to be any major real differences between the positions in the validation question reported in the literature. The majority of writers emphasize—in conformity with the extract given below—the empirical part of the model: "The adequacy of a model is determined by the nature of the state histories it produces. In broadest terms, the adequacy of a model is tested as follows: State histories produced by the model are compared with corresponding state histories produced by the modeled system"(Evans et al. 1967, pp. 8–9).

As regards the use of statistical methods in the testing of simulation models, most authors content themselves with listing and describing a number of statistical techniques that have relevance. Mention is made of, inter alia, factor, regression, spectral and variance analysis, Chi-square, Kolmogorov-Smirnov and non-parametric tests. Also mentioned is the predictive test "Theil's inequality coefficient" compiled in econometrics (Theil 1958). To this list we would add the predictive test "the Janus coefficient" by Gadd and Wold, which is introduced in Wold (1964).

Writers that have given attention to this and similar framings of the question are, inter alia, Cyert (1966), Fishman & Kiviat (1967*a–b*) and Naylor (1971).

In addition, a number of other ways of investigating models, based on general and subjective considerations of agreement, are reported; see e.g. Amstutz (1967). The Turing tests described in section 5.2.2 constitute one form of such validation. However, in the majority of cases these ways of posing the question have no quantitative alignment and are not intended for statistical and numerical treatment.

In conclusion, we make the following three general comments as regards the empirical validation of simulation models.

In the first place, we find that we get testing situations which are in principle no different from testing parametric hypotheses as dealt with in statistical theory.

In the second place, there has as far as I am aware not earlier been a detailed treatment of testing simulation models in the literature using systematically on the one hand hypothetical or population quantities and on the other hand estimated or sample quantities. An omission of the sample-population distinction for some entities tends to blur the problem of testing the models at some specified significance level. As a case in point: A correct model in all respects tested might be rejected only because its parameters are estimated and not equal to the population values if one does not make adjustment of the tests with the effect of the propagated error (5.2.19) or its equivalent.

And in the third place, empirical validation is just one—although important—step in the process of validating a simulation model intended for explanation. This point is supported by Meissner (1970), who in his simulation essay marks disagreements with writers such as e.g. Churchman (1963), who only stresses behavioural (prediction) aspects of simulation.

5.3. Testing the theoretical part of models

In this section the testing design presented falls within the frame of views of the testing of the theoretical part of simulation models, i.e. the study of correspondences between relations included in or excluded from the structure of the model and theoretical systems of hypotheses.

It is here first pointed out that the testing of the empirical part of a model naturally gives, even if indirectly, some knowledge of the theoretical part of the model, inasmuch as the formulation of the model is based upon a number of theoretical hypotheses. It is these hypotheses, translated into a model language and adapted to a real system, whose effects and consequences are controlled in the course of empirical testing.

As regards the direct testing of the theoretical part, no detailed treatment of this problem is, to my knowledge, given in the literature. Naylor and Finger (1967) appear to be the first to have dealt with the problem. Under the general heading "multi-stage validation" they have collected three principles in connection with an account of the views of three different schools in the philosophy of science, viz, Rationalism, Empiricism and Positive Economics. Their principles are intended to cover validation of both the theoretical and the empirical parts in a model.[1] The principles given in connection with "Rationalism" are in this respect an expression for the desideratum that the theoretical part in a model ought, briefly, to correspond to the accumulated state of theoretical knowledge concerning the modelled system. The way in which these correspondences are more closely evaluated in a constructed model is not, however, discussed in detail by them.

In the light of the model concept advanced in Chapter 2 we make the following observations concerning this problem:

(i) A certain kind of subjective validation of the theoretical part obtains in connection with the construction of the model, since there is a continous process of comparison with existing previous (theoretical) knowledge. In most cases this previous knowledge is reported by reference to parallel scientific works.

(ii) In a direct investigation of the model's premises in relation to theory it is difficult to arrive at acceptable criteria. Moreover, the method of controlling only scientific works used and reported in the model is sometimes insufficient. That this type of validation does not even occur in all circumstances appears from the fact that in the case of completely original innovations there are obviously no references for the basic approaches.

(iii) In an investigation of the conclusions drawn from the model there is the possibility of comparing "gross predictions" obtained with available theoretical knowledge which is not used in the model but has the same real referent; cf. the synthetic aspects of simulation given in section 2.4.

5.4. Testing of general models

In the presentation in the foregoing, models for particular real systems have been considered. Apart from the questions dealt with in section 4.2 on econometrics and on general systems theory in connection with the introduction of the general model-structure S, the idea of models intended for a class of not completely identical real systems has not been analysed in any detail. It is evident that special problems will arise in connection with the testing of such general models; cf. Bråten (1970*b*, 1971).

For the solution of the testing problem referring to the empirical part

[1] The concepts Rationalism, Empiricism and Positive Economics as interpreted by Naylor and Finger seem to have their equivalents in the concepts construct, content and predictive validation respectively in behavioural research; see e.g. Kerlinger (1964).

of general models, one cannot successfully make direct use of the specification advanced in section 4.4.

Thus a plausible answer to validity questions based on the definition of the model is that the model needs to be investigated for a sample of real systems with respect to its properties (Bråten & Norlén 1969), a requirement that in this formulation leads to a situation relatively difficult to master. In the treatment of general models interest should therefore, it would seem, in the first place and with advantage be focussed on direct investigations of the theoretical part of the models.

Chapter 6

Analysis of simulation models

6.1. Introduction

In this chapter the argument will concern problems arising in connection with the use of simulation experiments to establish quantitative model properties.

The numerical method is not a new construction; it has an established position as a recognized statistical method. By way of exemplification it may be mentioned that to help in deducing the distribution of Students' t-statistic, arranged observations of the statistic were used at an early stage (Student 1908).

This kind of numerical analysis plays an important role for the work and for the assessment of the model under consideration, and the field is therefore well suited for study and for penetrating critical attention. Here, too, of course, analytic deductions play a certain role, though they must be restricted in their application chiefly to parts of the model, e.g. to the analysis of subsystems to the model.

From the point of view of the experimenter the model system is in many respects easy to manage. The experimental system may be regarded as closed, and it permits of complete control at the same time as it makes possible reproduction of the results by replications of the experiment. From experiments with the system it is possible to estimate the effects (responses) of many influencing factors (stimuli). Factors are chosen in accordance with the target set for the experiment, and a set of treatment combinations is set up in the experimental design for the study of effects which are of interest.

Simulation experiments with models thus differ from experiments carried out in real systems. In experiments in real systems recourse is had to randomization in R. A. Fisher's sense of studied treatments of experimental units in order to get non-controllable experimental factors subjected to statistical control. In the case of simulation, on the other hand, all factors are under control and randomization need not be resorted to; the variation in the result as between different replications is incurred owing to the fact that different sequences of random numbers are used in the different replications.[1]

Below follows a discussion from the following list of five different aims in simulation experiments with the model:

(i) logical consistency in the model
(ii) description of the model
(iii) the stochastic properties of the model
(iv) sensitivity of the model
(v) the prediction properties of the model.

(i) *Logical consistency in the model.* The question as to whether the system constructed forms a logically coherent system arises immediately after the first construction performed. This constitutes the consistency aspect of the model. In the first experimental design the investigation may suitably be combined with the following aim.

(ii) *Description of the model.* The model is complex, and this makes it difficult or quite impossible purely in general to deduce implications with the help of simplifying deductions, and it reveals problems that add to the difficulty of understanding the model. There is therefore good reason to undertake simulations in order to procure the requisite amount of knowledge concerning the behaviour of the model. This is a necessary condition for the continued work of adaptation and testing. The use of descriptive statistics in the presentation of the model's behaviour plays, in this connection, an important role. The observation of the system added to observations of subsystems over longer simulated periods of time creates a need for data reduction and appurtenant illustrative presentations in a summarizing way. In the course of the work these problems will increasingly occupy the focus of attention, as it is frequently necessary to carry out the analyses on non-stationary behaviour.

(iii) *The stochastic properties of the model.* The stochastic nature of the model makes it necessary—for the determination of probable outcomes—to be able to assess the value of an individual observation on the model's behaviour. There arises, consequently, the need for analyses that report and give an account of the distribution of the disturbances in the model's behaviour for estimated parameters and values for predetermined variables. It is also of importance to find out how the distribution of the disturbances develops in a neighbourhood of these estimates.

In the previous chapter, in connection with the search for validity, we

[1] In this review real experiments in conformity with the classical definition i.e. with the possibility of independent replications under constant conditions, are considered. One may expect even larger differences between simulation experiments and real experiments when for the real experiments not all the conditions in the classical definition are completely met, i.e. in the intermediate cases in the transition from experimental to non-experimental situations.

have already given an account of the characteristic measurements that have been found to be central in the model. Thus with reference to representation (5.2.18) of the mean behaviour we stress once more the need to calculate the mean and the dispersion matrix of the propagated error (5.2.19) in the model. In addition to this we need an estimate of the dispersion matrices of the genuine variability of the model: first $D(\boldsymbol{\epsilon}_M)$ from (5.2.13–14) for estimated model, then $D(\tilde{\boldsymbol{\epsilon}}_M)$ from (5.2.15–16) for model with correct values for parameters and predetermined variables.

(iv) *Sensitivity of the model.* Those parts of the model that are robust with respect to changes in the parameters and predetermined variables of the model are of course easier to handle. These parts can be described with few statistics, and it should also be possible to reduce them in a simple way in eventual subsequent revisions of the model. Likewise, on the one hand the empirical measurements of these stable parts of the model may be permitted to have a low degree of precision, which means that the framing and estimation of characteristics will entail but little work. On the other hand, the parts that are sensitive to changes in the estimates of the model must be treated a good deal more carefully. This reasoning leads to the conclusion that a sensitivity analysis should be carried out, and this may be expected to result in rankings and in the best case even an estimate of the magnitude of the changes in the model's behaviour deriving from changes in the estimated parts of the model.

The above targets for the requisite sensitivity analysis indicated can, moreover, suitably be brought together with the already mentioned part-target constituted by estimation of the propagated error (5.2.19) according to the formulation in the previous chapter.

(v) *Prediction properties of the model.* Finally, mention is made of the need to procure observations of model behaviour for summarizing tests of the model and prediction concerning applications of the model. The operation O_M mentioned in the previous chapter is a description of the need for testing purposes. As regards the prediction aspect, this was discussed generally in Chapter 3.

Against the background of the above reasoning the reader is referred, as regards the first two points (i) and (ii), to Chapter 9 and to the Appendix, in which, with accompanying summaries, an account is given of simulations.

The next section, 6.2, gives a foundation for subsequent sections. In the continuation of the chapter the presentation has been restricted to a reference chiefly to the two pronouncedly analytic points (iii) and (iv) in the above list: the stochastic properties being briefly dealt with in section 6.3 and the sensitivity of the model being treated in section 6.4. The chapter ends with a largely expository section, 6.5, giving an account of experimental designs with reference to the presented needs.

6.2. Points of departure

The targets for analysis in simulation with the model may in general be said to be the elucidation of aspects of the implicitly given functions (4.2.35–38) or, which is the same thing, the functions (5.2.13–16).

From the experimental point of view the different kinds of parameters and predetermined variables entering in these relations are of equal value, and they are therefore only referred to as factors (stimuli) and are summarizingly designated in the form of the vector $\mathbf{x}$ with k elements. The estimate of $\mathbf{x}$ in the model is called the centre point and is designated with $\mathbf{x}^0$. The correct value of $\mathbf{x}$ is designated with $\boldsymbol{\xi}^0$. In this formulation the experiment generally involves both qualitative and quantitative variables. Only experiments with quantitative variables at interval-scale level will be discussed here.

Further, the vector $\boldsymbol{\eta}$ with n elements may be taken to designate the set of state variables (effects, responses) discussed in section 5.2. The relation system corresponding to (5.2.13–14) is written

$$\boldsymbol{\eta} = \boldsymbol{\phi}(\mathbf{x}) + \boldsymbol{\epsilon}(\mathbf{x}), \tag{6.2.1}$$

where

$$E(\boldsymbol{\eta} \mid \mathbf{x}) = \boldsymbol{\phi}(\mathbf{x}). \tag{6.2.2}$$

The analysis may be designated as a study of the system (6.2.1–2), first in the centre point $\mathbf{x}^0$ and second in a neighbourhood of this which with some confidence may be taken as containing the correct value $\boldsymbol{\xi}^0$.

Let us assume that altogether N experiments are carried out with the system. Let the matrix

$$\mathbf{X} = (x_{rs})_{N \times k} \tag{6.2.3}$$

designate the experimental design, where x_{rs} is the treatment of factor number s, measured from x_s^0, in experiment number r. Thus every row in $\mathbf{X}$ describes a treatment combination constituting values for the k factors. Let us assume further that the factor values have the mean $\mathbf{x}^0$, i.e. the column sums of (6.2.3) are all equal to 0. This can be written as

$$\boldsymbol{\iota}'\mathbf{X} = \mathbf{0}, \tag{6.2.4}$$

where $\boldsymbol{\iota}$ is the vector of the order N consisting of ones.

Let the matrix

$$\mathbf{Y} = (y_{rs})_{N \times n} \tag{6.2.5}$$

designate the effects from the N experiments, where y_{rs} indicates the value of variable number s in experiment number r.

Let us assume further that the function $\boldsymbol{\phi}$ for the values of $\mathbf{x}$ studied in a neighbourhood of the centre point $\mathbf{x}^0$ can be approximated with a good degree of accuracy by the first two terms in the Taylor development

$$\boldsymbol{\phi}(\mathbf{x}) \sim \boldsymbol{\phi}(\mathbf{x}^0) + \left[\frac{\partial \boldsymbol{\phi}}{\partial \mathbf{x}}\right]_{\mathbf{x}=\mathbf{x}^0} (\mathbf{x}-\mathbf{x}^0) = \boldsymbol{\mu} + \mathsf{B}(\mathbf{x}-\mathbf{x}^0), \tag{6.2.6}$$

which holds good if the function $\boldsymbol{\phi}$ and its first derivatives are continuous and if the neighbourhood of $\mathbf{x}^0$ is sufficiently small.[1] The stochastic system (6.2.1–2) may therewith be written

$$\boldsymbol{\eta} = \boldsymbol{\mu} + \mathsf{B}(\mathbf{x}-\mathbf{x}^0) + \boldsymbol{\epsilon}(\mathbf{x}, \mathbf{x}^0), \tag{6.2.7}$$

where

$$E(\boldsymbol{\eta} \,|\, \mathbf{x}, \mathbf{x}^0) = \boldsymbol{\mu} + \mathsf{B}(\mathbf{x}-\mathbf{x}^0) \tag{6.2.8}$$

for the values of $\mathbf{x}$ studied.

Then we can write

$$\underset{N\times n}{\mathbf{Y}} = \underset{N\times 1}{\boldsymbol{\iota}}\ \underset{1\times n}{\boldsymbol{\mu}'} + \underset{N\times k}{\mathbf{X}}\ \underset{k\times n}{\mathsf{B}'} + \underset{N\times n}{\mathbf{E}} \tag{6.2.9}$$

with

$$E(\mathbf{Y} \,|\, \mathbf{X}) = \boldsymbol{\iota}\boldsymbol{\mu}' + \mathbf{X}\mathsf{B}'. \tag{6.2.10}$$

The method of least squares applied relation for relation gives the following unbiassed estimators for the parameters in (6.2.9–10)

$$[\mathbf{c} \,\vdots\, \mathbf{B}] = \mathbf{Y}'[\boldsymbol{\iota} \,\vdots\, \mathbf{X}] \begin{bmatrix} \boldsymbol{\iota}'\boldsymbol{\iota} & \boldsymbol{\iota}'\mathbf{X} \\ \mathbf{X}'\boldsymbol{\iota} & \mathbf{X}'\mathbf{X} \end{bmatrix}^{-1}. \tag{6.2.11}$$

The assumption (6.2.4) and $\boldsymbol{\iota}'\boldsymbol{\iota} = N$ permit us to write the estimator also as follows

$$[\mathbf{c} \,\vdots\, \mathbf{B}] = \left[\frac{1}{N}\mathbf{Y}'\boldsymbol{\iota} \,\vdots\, \mathbf{Y}'\mathbf{X}(\mathbf{X}'\mathbf{X})^{-1}\right]. \tag{6.2.12}$$

If the distribution for the disturbance $\boldsymbol{\epsilon}$ is normal and independent of $\mathbf{x}$, then it can be shown that the estimator is also the maximum-likelihood estimator.[2]

[1] In general terms we conclude that the neighbourhood of $\mathbf{x}^\circ$ may be made quite small. The difference $\mathbf{x}^\circ - \boldsymbol{\xi}^\circ$ in general consists of both sampling and non-sampling errors. Consider the case when all the factors $\mathbf{x}$ are model parameters. Furthermore, assume that the estimates $\mathbf{x}^\circ$ are obtained through the application of some "optimal" estimation method from independent observations without measurement errors (section 7.3). It then follows that the magnitude of the difference $\mathbf{x}^\circ - \boldsymbol{\xi}^\circ$ as the length of confidence intervals will be of the order $N^{-\frac{1}{2}}$, where N is the number of observations. This comment thus supports the use of the Taylor expansion (6.2.6).

[2] See section 7.3 for a brief summary of least-squares and maximum likelihood estimation.

6.3. Analysis of stochastic properties

As regards stochastic properties of a model, the points we wish to make at this stage are the following.

From the relation (6.2.9) it follows that the matrix with disturbances can be estimated with

$$\mathbf{Y}-\boldsymbol{\iota}\mathbf{c}'-\mathbf{X}\mathbf{B}', \tag{6.3.1}$$

where the estimates $\mathbf{c}$ and $\mathbf{B}$ are obtained from (6.2.12). These residuals have the mean value zero, as may be seen from the premultiplication of (6.3.1) by $\boldsymbol{\iota}'$

$$\boldsymbol{\iota}'\mathbf{Y}-\boldsymbol{\iota}'\boldsymbol{\iota}\mathbf{c}'-\boldsymbol{\iota}'\mathbf{X}\mathbf{B}'=N\left(\frac{1}{N}\mathbf{Y}'\boldsymbol{\iota}-\mathbf{c}\right)'=\mathbf{0}'. \tag{6.3.2}$$

If the distribution of the disturbances is independent of $\mathbf{x}$, then it can be shown that an unbiassed estimator of the dispersion matrix is

$$\mathbf{S}_M=\frac{1}{N-k-1}\mathbf{Y}'\left(\mathbf{I}-\frac{1}{N}\boldsymbol{\iota}\boldsymbol{\iota}'-\mathbf{X}(\mathbf{X}'\mathbf{X})^{-1}\mathbf{X}'\right)\mathbf{Y}, \tag{6.3.3}$$

which may thus be used as estimator of the dispersion matrices for both $\boldsymbol{\epsilon}_M$ in (5.2.13) and $\tilde{\boldsymbol{\epsilon}}_M$ in (5.2.15).

It is of interest to investigate the plausibility of this assumption that the dispersion matrix for $\boldsymbol{\epsilon}_M$ may be regarded as equal to the dispersion matrix of $\tilde{\boldsymbol{\epsilon}}_M$. To this end the effect matrix $\mathbf{Y}$ and the observation matrix $\mathbf{X}$ can be arranged and partitioned according to

$$\mathbf{Y}=\begin{bmatrix}\underset{N_1\times n}{\mathbf{Y}_1}\\ \hdashline \underset{N_2\times n}{\mathbf{Y}_2}\end{bmatrix};\quad X=\begin{bmatrix}\underset{N_1\times k}{\mathbf{X}_1}\\ \hdashline \underset{N_2\times k}{\mathbf{X}_2}\end{bmatrix} \tag{6.3.4}$$

so that $\mathbf{Y}_1$ is equal to the effects in experiments with $\mathbf{X}_1=\mathbf{0}$. Then form the following estimators

$$\mathbf{S}_\varepsilon=\frac{1}{N_1-1}\mathbf{Y}_1'\left(\mathbf{I}-\frac{1}{N_1}\boldsymbol{\iota}\boldsymbol{\iota}'\right)\mathbf{Y}_1 \tag{6.3.5}$$

$$\mathbf{S}_{\tilde{\varepsilon}}=\frac{1}{N_2-k-1}\mathbf{Y}_2'\left(\mathbf{I}-\frac{1}{N_2}\boldsymbol{\iota}\boldsymbol{\iota}'-\mathbf{X}_2(\mathbf{X}_2'\mathbf{X}_2)^{-1}\mathbf{X}_2'\right)\mathbf{Y}_2 \tag{6.3.6}$$

for $D(\boldsymbol{\epsilon}_M)$ and $D(\tilde{\boldsymbol{\epsilon}}_M)$ respectively. It is possible to apply a test then of whether a similarity between the dispersion matrices can be considered to exist. If the null hypothesis is rejected it follows that (6.3.3) cannot be used as a common estimator for $D(\boldsymbol{\epsilon}_M)$ and $D(\tilde{\boldsymbol{\epsilon}}_M)$. The estimator (6.3.5) may still of course be used for $D(\boldsymbol{\epsilon}_M)$, and the problem arising consists in finding an estimator for $D(\tilde{\boldsymbol{\epsilon}}_M)$.

6.4. Sensitivity analysis

It may be as well to preface the discussion of the sensitivity of the system with a study of the propagated error (5.2.19) and its dependence on the factor estimate $\mathbf{x}^0$. The use of (6.2.8) gives

$$\boldsymbol{\epsilon}_\phi = E(\boldsymbol{\eta}\,|\,\mathbf{x}^0, \mathbf{x}^0) - E(\boldsymbol{\eta}\,|\,\boldsymbol{\xi}^0, \mathbf{x}^0) = \mathsf{B}(\mathbf{x}^0 - \boldsymbol{\xi}^0). \tag{6.4.1}$$

If $\mathbf{x}^0$ is an unbiassed estimator of $\boldsymbol{\xi}^0$ then the expectation of the propagated error is

$$E(\boldsymbol{\epsilon}_\phi) = E\,\mathsf{B}(\mathbf{x}^0 - \boldsymbol{\xi}^0) = \mathsf{B}(\boldsymbol{\xi}^0 - \boldsymbol{\xi}^0) = \mathbf{0}, \tag{6.4.2}$$

which thus simplifies the calculation of e.g. the distribution of the variable (5.2.29). The dispersion matrix for the propagated error is obtained as

$$D(\boldsymbol{\epsilon}_\phi) = \mathsf{B}D(\mathbf{x}^0)\,\mathsf{B}'. \tag{6.4.3}$$

If the analysis of the real system has given the estimate S_R of $D(\mathbf{x}^0)$, it follows that

$$\mathbf{S}_\phi = \mathbf{B}\mathbf{S}_R\mathbf{B}' \tag{6.4.4}$$

can be used as an estimate of (6.4.3) in, inter alia, a calculation of the distribution of (5.2.29), where $\mathbf{B}$ is obtained from (6.2.12).

Besides sensitivity in the above-described sense there is the earlier need, referred to in the introduction, of a general study of the function $\boldsymbol{\phi}$ and the disturbance $\boldsymbol{\epsilon}$ in (6.2.1–2). If we bear in mind the approximation (6.2.8–9) it will be realized that the estimate $\mathbf{B}$ from (6.2.13) and the analysis of the stochastic properties described in the previous section are relevant also in this context.

In the sequel we shall deal only with the effects depending upon the form of the function $\boldsymbol{\phi}$. In order to obtain an estimate of sensitivity that is somewhat independent of the scales used we shall define sensitivity with

Definition 6.1. *For the system (6.2.1–2) the sensitivity of η_i with respect to x_j in the point $\mathbf{x}$ is*

$$\nu_{ij}(\mathbf{x}) = \frac{\partial \log \phi_i(\mathbf{x})}{\partial \log x_j} = \frac{\partial \phi_i(\mathbf{x})}{\partial x_j} \cdot \frac{x_j}{\phi_i(\mathbf{x})}. \tag{6.4.5}$$

The above definition is an extension of the concept of sensitivity as this is defined for deterministic systems in the theory for feedback and control systems, see e.g. Di Stefano III et al. (1967). In this connection it is also of interest to note the similarity between the above definition and the concept of elasticity in economics introduced by Marshall (1890).

The sensitivity—or elasticity, as (6.4.5) may also be called—is approximatively

$$\nu_{ij}(\mathbf{x}) \sim \frac{\phi_i(\mathbf{x}+\boldsymbol{\Delta}_j) - \phi_i(\mathbf{x})}{\phi_i(\mathbf{x})} \Bigg/ \frac{x_j + \Delta_j - x_j}{x_j}, \tag{6.4.6}$$

where Δ_j is a small number and where

$$\mathbf{x}+\boldsymbol{\Delta}_j = \begin{bmatrix} x_1 \\ \vdots \\ x_j \\ x_j+\Delta_j \\ x_{j+1} \\ \vdots \\ x_k \end{bmatrix}. \tag{6.4.7}$$

The sensitivity may thus be expressed as the quotient between the changes in the function ϕ_i and the factor x_j. Let us assume that the value for the factor is increased by 1 per cent, i.e. Δ_j is equal to 0.01 x_j. Then the sensitivity v_{ij} will indicate approximatively by what percentage the function ϕ_i increases or diminishes.

Write the ith row in (6.2.6) as

$$\phi_i(\mathbf{x}) = \mu_i + \sum_{j=1}^{k} \beta_{ij}(x_j - x_j^0). \tag{6.4.8}$$

Hence

$$\phi_i(\mathbf{x}^0) = \mu_i \tag{6.4.9}$$

$$\frac{\partial \phi_i(\mathbf{x})}{\partial x_j} = \beta_{ij} \tag{6.4.10}$$

and as an estimate of the sensitivity of η_i with respect to x_j in the point $\mathbf{x}^0$ one may thus use

$$v_{ij}(\mathbf{x}^0) = \frac{b_{ij} \cdot x_j^0}{c_i}, \tag{6.4.11}$$

where b_{ij} and c_i are obtained from (6.2.12).

As emerges from the foregoing, however, the above measure of the sensitivity gives only a partial solution to the problem of measuring the total changes in effect that follow when the values for the factors are changed. Thus one does not get an immediate answer to questions concerning the dynamic implications in the behaviour of the system, how the generated time-series for e.g. the state variables change. In this connection the following is suggested.

The conditional expectation formed in the difference equation (4.2.36) constitutes the point of departure. The factor $\mathbf{x}$ for which the sensitivity is studied consists of a subset of initial states $\boldsymbol{\eta}_0$, parameters $\boldsymbol{\alpha}$ and input variables ${}^0\boldsymbol{\zeta}_\tau$ in the equation mentioned.

As in the case of the earlier formulation of the problem, we study the sensitivity for the factor $\mathbf{x}$, whose values are varied in a neighbourhood of the centre point $\mathbf{x}^0$.

The system may be said to have a low degree of sensitivity with respect to $\mathbf{x}$ if the solution to (4.2.36) for $\mathbf{x}=\mathbf{x}^0$ is practically equal to the solutions for the values of $\mathbf{x}$ in the vicinity of $\mathbf{x}^0$. If, on the other hand, the solutions vary considerably, this shows that the system is sensitive with respect to $\mathbf{x}$.

A mathematical treatment of the sensitivity may be carried out as follows. One can introduce small changes for the components corresponding to $\mathbf{x}$, firstly in the parameter vector with

$$\mathbf{a}^{\Delta} = \mathbf{a} - \mathbf{a}^0 \tag{6.4.12}$$

and secondly in the time-series of inputs with

$$\mathbf{z}_{\tau}^{\Delta} = \mathbf{z}_{\tau} - {}^0\mathbf{z}_{\tau}, \tag{6.4.13}$$

where $\mathbf{a}^0$ and ${}^0\mathbf{z}_{\tau}$ are the estimates in the model. The differential of (4.2.36) is written

$$dE(\boldsymbol{\eta}_{\tau+1} \mid \mathbf{a}, \boldsymbol{\eta}_{\tau}, \mathbf{z}_{\tau}) = \frac{\partial \boldsymbol{\varkappa}}{\partial \mathbf{a}} d\mathbf{a} + \frac{\partial \boldsymbol{\varkappa}}{\partial \boldsymbol{\eta}_{\tau}} d\boldsymbol{\eta}_{\tau} + \frac{\partial \boldsymbol{\varkappa}}{\partial \mathbf{z}_{\tau}} d\mathbf{z}_{\tau}. \tag{6.4.14}$$

An approximation analogous with (6.2.6) gives

$$\boldsymbol{\eta}_{\tau+1}^{\Delta} = \mathsf{A}\mathbf{a}^{\Delta} + \mathsf{B}\boldsymbol{\eta}_{\tau}^{\Delta} + \boldsymbol{\Gamma}\mathbf{z}_{\tau}^{\Delta} + \boldsymbol{\epsilon}_{\tau+1}^{\Delta} \tag{6.4.15}$$

with

$$E(\boldsymbol{\eta}_{\tau+1}^{\Delta} \mid \mathbf{a}^{\Delta}, \boldsymbol{\eta}_{\tau}^{\Delta}, \mathbf{z}_{\tau}^{\Delta}) = \mathsf{A}\mathbf{a} \quad + \mathsf{B}\boldsymbol{\eta}_{\tau}^{\Delta} + \boldsymbol{\Gamma}\mathbf{z}_{\tau}^{\Delta}, \tag{6.4.16}$$

where

$$\boldsymbol{\eta}_{\tau}^{\Delta} = \boldsymbol{\eta}_{\tau} - \boldsymbol{\eta}_{\tau}^{0} \tag{6.4.17}$$

for $\tau = 0$ represents the change in the initial values $\boldsymbol{\eta}_0$ and for $\tau > 0$ represents the deviation from the reference solution $\boldsymbol{\eta}_{\tau}^{0}$, which is obtained from small changes (6.4.12) in the parameter values and small changes (6.4.13) in the time-series of inputs. Further, the coefficient matrices A, B and $\boldsymbol{\Gamma}$ give the value of the partial derivatives which are assumed to be constant.

The system (6.4.15–16) is a stochastic counterpart of the deterministically given variational equation in control theory. In this case the nearest counterpart of the variational equation in control theory is the relation (6.4.16).[1]

Estimates of the coefficient matrices are obtained either from experimental results or from (6.4.14) from an earlier reduction of the system to a relation system.

For the further treatment of the system (6.4.15–16) and the establishing of its characteristic properties with the help of, inter alia, the so-called z-transform the reader is referred to other works; see e.g. Robinson (1967) and Tou (1964).

6.5. Factor values

For a general introduction to the design and analysis of experiments the reader is referred to standard works by, inter alios, Fisher (1935), Cox (1958) and Davies (1967). In this section we shall discuss briefly the choice of experimental points, i.e. the choice of factor values in the matrix $\mathbf{X}$ with reference to the estimation requirements emerged.

An inventory of the restrictions on the design-matrix so far imposed gives

$\boldsymbol{\iota}'\mathbf{X} = \mathbf{0}$ the mean for the factor values is equal to $\mathbf{x}_0$.
$|\mathbf{X}'\mathbf{X}| \neq 0$ the dispersion-matrix for the factor values is not singular.
$N_1 \geqslant 2$ the number of experiments with the factor values equal to $\mathbf{x}^0$ is greater than or equal to 2.
$N_2 \geqslant k+2$ the number of experiments with the factor values different from $\mathbf{x}^0$ is greater than or equal to the number of factors plus two.

A study of the estimators in the previous section first shows the necessity of obtaining estimates $\mathbf{c}$ and $\mathbf{B}$ according to (6.2.12) and the estimates v in (6.4.11) based thereon. The variances and covariances for $\mathbf{c}$ and $\mathbf{B}$ consist of the elements in the square matrix

$$D(\boldsymbol{\epsilon}) \otimes \begin{bmatrix} 1/N & \mathbf{0}' \\ \mathbf{0} & (\mathbf{X}'\mathbf{X})^{-1} \end{bmatrix} \tag{6.5.1}$$

of the order $n(k+1)$ that is obtained from Kronecker or the direct product of the two matrices entering if the distribution of the disturbances in (6.2.7) is assumed to be independent of $\mathbf{x}$.

The studied neighbourhood of $\mathbf{x}^0$ is represented by the interval

$$-\boldsymbol{\theta} \leqslant \mathbf{x} - \mathbf{x}^0 \leqslant \boldsymbol{\theta}, \tag{6.5.2}$$

where $\boldsymbol{\theta}$ is a vector with positive constants. The variances in (6.5.1) are minimized if the experimental points not equal to the centre point $\mathbf{x}^0$ are located at the terminal points of the interval.

A simple design having certain optimal properties is thus constituted by $N = N_1 + N_2$ experiments, where

N_1 = the number of centre points $\mathbf{x} = \mathbf{x}^0$.
$N_2 = 2\,km$ axis points, which for every $i = 1, ..., k$ consists of on one hand m points

$$x_j = x_j^0 + \begin{cases} \theta_i \text{ for } j = i \\ 0 \text{ for } j \neq i \end{cases} \quad j = 1, ..., k \tag{6.5.3}$$

---

[1] The variational equation is defined in Šiljak (1969). A survey is given by Kokotović & Rutman (1965).

and on the other hand m points

$$x_j = x_j^0 + \begin{cases} -\theta_i \text{ for } j=i \\ 0 \text{ for } j \neq i \end{cases} \quad j=1, \ldots, k \tag{6.5.4}$$

For this design we get the following dispersion matrix for the factor values

$$\mathbf{X}'\mathbf{X} = \frac{N_2}{N}\mathbf{\Theta} \tag{6.5.5}$$

where $\mathbf{\Theta}$ is the diagonal matrix with elements θ_i^2.

What is decisive for the analyses is the validity of the linear approximation in (6.2.6). From the combined centre- and axis-design above it is possible to carry out tests of the linearity assumption. If it proves that the linear approximation is not suitable then—if suitable continuity properties are fulfilled—every element $\phi_i(\mathbf{x})$ in the vector function $\phi(\mathbf{x})$ can be approximated with more accuracy than before by

$$\phi_i(\mathbf{x}) \sim \phi_i(\mathbf{x}^0) + \left(\frac{\partial \phi_i(\mathbf{x})}{\partial \mathbf{x}}\right)'_{\mathbf{x}=\mathbf{x}^0} (\mathbf{x}-\mathbf{x}^0) + \tfrac{1}{2}(\mathbf{x}-\mathbf{x}^0)' \left[\frac{\partial^2 \phi_i(\mathbf{x})}{\partial \mathbf{x}^2}\right]_{\mathbf{x}=\mathbf{x}^0} (\mathbf{x}-\mathbf{x}^0)$$

$$= \mu_i + \boldsymbol{\beta}_i'(\mathbf{x}-\mathbf{x}^0) + (\mathbf{x}-\mathbf{x}^0)'\mathbf{\Gamma}_i(\mathbf{x}-\mathbf{x}^0). \tag{6.5.6}$$

Thus every row in (6.2.1–2) may be written

$$\eta_i = \mu_i + \boldsymbol{\beta}_i'(\mathbf{x}-\mathbf{x}^0) + (\mathbf{x}-\mathbf{x}^0)'\mathbf{\Gamma}_i(\mathbf{x}-\mathbf{x}^0) + \varepsilon_i(\mathbf{x}, \mathbf{x}^0) \tag{6.5.7}$$

and

$$E(\eta_i \,|\, \mathbf{x}, \mathbf{x}^0) = \mu_i + \boldsymbol{\beta}_i'(\mathbf{x}-\mathbf{x}^0) + (\mathbf{x}-\mathbf{x}^0)'\mathbf{\Gamma}_i(\mathbf{x}-\mathbf{x}^0) \tag{6.5.8}$$

respectively. In the terminology of variance analysis the system (6.2.7–8), on the one hand, permits of an estimation of the mean effects of the factors; the system (6.5.7–8), on the other hand, permits also of an estimation of interaction effects of the second order from the factors.

For an estimation of the parameters in the system (6.5.7–8) the combined centre and axis design indicated is not sufficient. We have the possibility of extending the experiments with the full factorial design with two values for each factor. This design consists of all combinations

$$\mathbf{x} = \mathbf{x}^0 + \begin{pmatrix} \pm\theta_1 \\ \vdots \\ \pm\theta_k \end{pmatrix} \tag{6.5.9}$$

of points in the factor space. The number of such points is

$$N_3 = 2^k, \tag{6.5.10}$$

which is thus great even with relatively few factors. Where there are many factors it is therefore suitable to use only one subset of the points (6.5.10) constituting a fraction 2^p of the full factorial design, in other words a fractional factorial design consisting of

$$N_3' = 2^{k-p} \tag{6.5.11}$$

experimental points. For surveys of factorial designs, see e.g. Box & Hunter (1961) and Draper & Stoneman (1966). Tables for such designs have been prepared (Statistical Engineering Laboratory of National Bureau of Standards 1957).

Theories for experiments in this and similar situations have been developed and reported in a number of works. In connection with the fractional factorial design mention may be made of works that have given an account of studies concerning the confounding of interaction effects in estimation that arises where the fraction 2^p is great; see e.g. Davies (1967). It should here be mentioned that the design indicated in the foregoing, with altogether

$$N = N_1 + N_2 + N_3(N_3') \tag{6.5.12}$$

points, is dealt with in another and summarizing way and to a certain extent also tabulated as an experiment with factors with more than two levels (Connor & Zelen 1959, Connor & Young 1961).

The choice of a design of course reflects the properties that have been considered to be most important in the experiment. Box (1964) has given a list of desirable properties for designs. As regards factorial and fractional factorial designs, these are often chosen on account of their simplicity. Two other recommended criteria are rotatability and the $|\mathbf{X}'\mathbf{X}|$-criterion. As regards the rotatability criterion reference is made to Cochran & Cox (1957). The second criterion may be described as maximation of the determinant value $|\tilde{\mathbf{X}}'\tilde{\mathbf{X}}|$, where

$$\tilde{\mathbf{X}} = \begin{bmatrix} \left(\dfrac{\partial\phi(\boldsymbol{\lambda}, \mathbf{x})}{\partial\boldsymbol{\lambda}}\right)'_{\mathbf{x}=\mathbf{x}_1} \\ \vdots \\ \left(\dfrac{\partial\phi(\boldsymbol{\lambda}, \mathbf{x})}{\partial\boldsymbol{\lambda}}\right)'_{\mathbf{x}=\mathbf{x}_N} \end{bmatrix} \tag{6.5.13}$$

and where $\boldsymbol{\lambda}$ is the parameter vector that is to be estimated in the function ϕ and $\mathbf{x}_1, \ldots, \mathbf{x}_N$ are the experimental points; see e.g. Box & Draper (1971). For a linear model approximation of ϕ according to (6.2.7–8) $\tilde{\mathbf{X}}$ in (6.5.13) thus agrees with (6.2.3).

In the following example a design is developed that proceeds from another criterion and is based on a specially indicated circumstance that occurs in experiments with simulation models. It concerns the bias of the estimator (6.4.4) of the dispersion matrix for the propagated error in the model.

Example 6.1. The estimator (6.4.4) of the dispersion matrix for the propagated error (5.2.19) is chosen as point of departure. First we shall show that this estimator is biassed. Then we derive an experimental design which minimizes this bias and outline a procedure for the generation of ensuing experimental points.

Using (6.2.9) and (6.2.12) the estimator may be written

$$\begin{aligned} \mathbf{S}_\phi &= \mathbf{B}\mathbf{S}_R\mathbf{B}' = \mathbf{Y}'\mathbf{X}(\mathbf{X}'\mathbf{X})^{-1}\mathbf{S}_R(\mathbf{X}'\mathbf{X})^{-1}\mathbf{X}'\mathbf{Y} \\ &= (\boldsymbol{\mu}\boldsymbol{\iota}' + \mathsf{B}\mathbf{X}' + \mathbf{E}')\mathbf{X}(\mathbf{X}'\mathbf{X})^{-1}\mathbf{S}_R(\mathbf{X}'\mathbf{X})^{-1}\mathbf{X}'(\boldsymbol{\iota}\boldsymbol{\mu}' + \mathbf{X}\mathsf{B}' + \mathbf{E}) \\ &= \mathsf{B}\mathbf{S}_R\mathsf{B}' + \mathbf{E}'\mathbf{X}(\mathbf{X}'\mathbf{X})^{-1}\mathbf{S}_R\mathsf{B}' + \mathsf{B}\mathbf{S}_R(\mathbf{X}'\mathbf{X})^{-1}\mathbf{X}'\mathbf{E} \\ &\quad + \mathbf{E}'\mathbf{X}(\mathbf{X}'\mathbf{X})^{-1}\mathbf{S}_R(\mathbf{X}'\mathbf{X})^{-1}\mathbf{X}'\mathbf{E}. \end{aligned} \tag{6.5.14}$$

For a given $\mathbf{S}_R$ we get the expectation of (6.5.14) as

$$E(\mathbf{S}_\phi \,|\, \mathbf{S}_R) = \mathsf{B}\mathbf{S}_R\mathsf{B} + \operatorname{tr}\{\mathbf{S}_R(\mathbf{X}'\mathbf{X})^{-1}\}\, D(\boldsymbol{\epsilon}) \tag{6.5.15}$$

if the distribution of the disturbances is independent of $\mathbf{x}$. The second term in the right side is thus the bias which arises as a consequence of the estimation of the parameter matrix B.

Write this term

$$\text{Bias} = \operatorname{tr}\{\mathbf{S}_R(\mathbf{X}'\mathbf{X})^{-1}\}\, D(\boldsymbol{\epsilon}) = \frac{1}{N}\operatorname{tr}\{\mathbf{S}_R\boldsymbol{\Sigma}_M^{-1}\}\, D(\boldsymbol{\epsilon}), \tag{6.5.16}$$

where N is the number of experimental points and $\boldsymbol{\Sigma}_M$ is the dispersion matrix of the factor values.

A small bias can thus be obtained with the help of a design in which the factor values are correlated with each other with a dispersion matrix that minimizes the bias.

For the design, consider the case in which the elements in $\mathbf{X}$ are constituted by the end-points of the interval (6.5.2), in which connection, for every factor i, $N/2$ values x_i are equal to $x_i^0 + \theta_i$ and $N/2$ values are equal to $x_i^0 - \theta_i$, and where N is assumed to be an even integer.

Thus

$$\operatorname{Diag}(\boldsymbol{\Sigma}_M) = \boldsymbol{\Theta}. \tag{6.5.17}$$

Consider the Lagrange function

$$f = \operatorname{tr}\{\mathbf{S}_R\boldsymbol{\Sigma}_M^{-1}\} + \operatorname{tr}\{\boldsymbol{\Lambda}(\boldsymbol{\Sigma}_M - \boldsymbol{\Theta})\}, \tag{6.5.18}$$

where $\boldsymbol{\Lambda}$ is a diagonal matrix that contains Lagrange multipliers.

If the derivatives of the first order of (6.5.18) are put equal to zero we get

$$\mathbf{S}_R = \mathbf{\Sigma}_M \mathbf{\Lambda} \mathbf{\Sigma}_M, \tag{6.5.19}$$

which together with the condition (6.5.17) gives a non-linear equation system with as many equations as there are unknowns.

Consider the simple case with two factors, where

$$\mathbf{S}_R = \begin{bmatrix} 1 & r \\ r & 1 \end{bmatrix}; \quad \mathbf{\Theta} = \begin{bmatrix} 1 & 0 \\ 0 & 1 \end{bmatrix}. \tag{6.5.20}$$

The equations (6.5.17) and (6.5.19) give the solution

$$\varrho = r/(1 + \sqrt{1 - r^2}). \tag{6.5.21}$$

The correlation ϱ in $\mathbf{\Sigma}_M$ is thus numerically less than the correlation r in $\mathbf{S}_R$. The bias (6.5.16) becomes $(1 + \sqrt{1 - r^2})\, D(\boldsymbol{\epsilon})/N$, which may be compared, for example, with a design where ϱ is equal to zero (e.g. a full factorial design) in which case the bias is $2D(\boldsymbol{\epsilon})/N$.

The design matrix $\mathbf{X}$ can be obtained in the following way. Generate observations $\boldsymbol{\xi}$ from the multivariate normal distribution $\mathrm{N}(\mathbf{0}, \mathbf{\Omega})$ with diag $(\mathbf{\Omega}) = \mathbf{I}$. For each factor i, put

$$x_i = x_i^0 \begin{cases} +\theta_i & \text{if} \quad \xi_i \geqslant 0 \\ -\theta_i & \text{if} \quad \xi_i < 0. \end{cases} \tag{6.5.22}$$

The correlation ϱ_{ij} between the factors x_i and x_j is therewith

$$\varrho_{ij} = \iint_{D_1} - \iint_{D_2} f(u, t)\, du\, dt, \tag{6.5.23}$$

where f is the bivariate normal distribution with the mean values 0, variances 1 and correlation coefficient ω_{ij}. The integration area D_1 is constituted by the 1st and 3rd quadrants and the area D_2 is constituted by the 2nd and 4th quadrants. Evaluation of the integral gives

$$\varrho_{ij} = \frac{2}{\pi} \arcsin(\omega_{ij}). \tag{6.5.24}$$

The correlations ω_{ij} in $\mathbf{\Omega}$ ought thus to be calculated from

$$\omega_{ij} = \sin\left(\frac{\pi}{2} \varrho_{ij}\right). \tag{6.5.25}$$

It should, however, be mentioned that not all designs can be got with the help of this technique, since (6.5.25) makes it clear that the absolute value of ω is greater than the absolute value of ϱ. A given matrix $\mathbf{\Sigma}_M$ may thus lead to an unrealizable matrix $\mathbf{\Omega}$.

The experimental points will have a mean value of $\mathbf{x}^0$ if instead of (6.5.22) we use

$$x_i = \begin{cases} +\theta_i \text{ if } \xi_i \geqslant 0 \text{ and fewer than } N/2 \text{ of the earlier generated} \\ \qquad \text{points } \xi_i \text{ have this value} \\ -\theta_i \text{ otherwise.} \end{cases} \tag{6.5.26}$$

In the sample, of course, the factor values obtained will in general have a dispersion matrix $\mathbf{S}_M$ that deviates from $\mathbf{\Sigma}_M$. If, however, a sufficiently large number of samples are drawn in the way indicated, then the "best" design will have a dispersion matrix that is in close agreement with $\mathbf{\Sigma}_M$.

As a final comment to this chapter concerning the design of simulation experiments we make the observation that the estimation of distribution properties of simulated model behaviour in general is a clearly specified technical problem. Observe, for example, the case where every simulation gives an observation from a statistical distribution. If statistically independent simulations are performed, one generally obtains estimates of distribution characteristics which have standard errors that are inversely proportional to $\sqrt{N}$, where N is the number of simulations (observations).[1] This is one reason why the problem of planning experimental designs with estimators that yield estimates having small standard errors has been given a relatively narrow treatment in this study. The so-called variance reduction techniques that are used here may in the light of the present discussion be said to mean smaller constants of proportionality when calculating the standard errors of the estimators.

[1] This argument is based on the theory of the sampling distribution for functions of moments as given by Cramér (1945).

Chapter 7

Analysis of modelled systems

7.1. Introduction

The analysis of modelled systems shows features which in several respects resemble those earlier described for the analysis of model systems. In both cases the primary aim is to put the researcher in observation contact with the system studied. In the analysis of the model system this contact could be established preferably in connection with the carrying out of an experimental design covering a series of simulations with the model on a computer. The analysis of the real system, on the other hand, is performed mainly on the basis of survey designs of the studied system consisting of plans for the collection, control and continued processing of observational data from the real system.

In comparison with the analysis of the model system the analysis of the real system is in several respects more circumstantial. The chief reason for this, as mentioned earlier, is that the structure of the real system is for the considered cases not observable. In cybernetic terminology the system is called a black box (Klir & Valach 1967). Observation of the behaviour of the system in this case takes as its point of departure an hypothesis of the structure of the system in the form of a model for the same. Analysis of the real system is thus in the first place carried out with reference to the model for the system.

Real systems are characterized by the fact that they are always more or less open. This means that during observation of the system many variables cannot be controlled. Furthermore, on account of these circumstances important variables are probably not even known or may be overlooked and not reported in the model.

Thus in this case of analysis of the real system an attempt ought to be made, in this non-experimental situation, to replace the previous possibility of experimental control of the model system with an ideal model allowing of the incorporation of possible and relevant factors in the real system. Success, in carrying out an analysis of the real system is, on the whole, dependent on these potentialities of the model, i.e. its ability to take relevant factors into account and its possibilities of considering eventual disturbances in the reporting of observations, which reflect the effect of factors of non-constant character during the observation period, which is not, either, of primary interest.

The discussion in Chapters 3–6 leads to the conclusion, deduced from a study of the different phases of the modelling work, that several real analyses are needed, incorporating also the preparatory studies preceding the first construction of the computer model. Several different needs for analysis and a number of problems arise in the course of the work, according to the particular stages of development of the latter. A first glance at the field discloses the following three prime requirements as regards extended knowledge of the real system.

For the construction and specification of the model: (i) description of the real system, (ii) exploration in the real system, and for the evaluation of the model (iii) estimation of model parameters and values of model variables.

We make the following observations as regards the itemized points.

(i) *Description of the real system.* In the first phase of the modelling work the self-evident need arises for descriptive investigations to gain information for continued work. Since during the working stage the model is in several respects not completely developed it is of course not possible to use it to show what records of observations should be made. In the majority of cases, however, the limits of the model are so far indicated that it is possible to make a first record of needed observations in the form of tentative catalogues of variables for objects in the system.

The descriptive investigations play an important role for the continued work of analysis, and in the respect referred to here this may be compared with the role that corresponding investigations play for the analysis of the model system described at the beginning of the previous chapter.

(ii) *Exploration in the real system.* Besides the general need for summarizing accounts there is the need of a broader basis for a more detailed penetration of problems and even of new experience concerning the ways in which the real system behaves.

Thus in the longer perspective of the search for targets one is able to distinguish targets in the form of expectations of further reportable basic data for a reinforcement of the foundations for the work, with the requisite specification of the model and its properties, including, inter alia, the formulation of hypotheses for the behaviour in the form of different model-mechanisms.

These penetrations of, or exploratory investigations of, the real system are of course supported by existing previous knowledge of the object studied. Several alternative and perhaps also competing explanations of the conditions studied may be predicted and formulated in hypotheses as to how the behaviour is constituted or composed. The function of the exploratory work in this connection is to ascertain the explanation or explanations that may be adjudged most reasonable as compared with the empirical evidence obtained. This formulation of the problem shows the goal for the analysis to be closely related to the next aim.

(iii) *Estimates of model parameters and values of model variables.* After the preparatory analyses (i) and (ii) have been carried out and after the construction of the model there arise the easily recognizable analytic problems with which one is confronted in the work connected with the estimation and testing of the model.

In this connection the specified model, as mentioned earlier, forms the point of departure for the determination of the types of observations needed and for the continued utilization of the observations obtained.

In Chapter 5 we have described the kinds of original and revised records of observations which—on the basis of the testing requirements for the model—need to be made; particular stress was laid on the need for estimates of model parameters and of subsequent testings of these estimates and estimates of values of predetermined variables. Estimates of values of variables in the real system corresponding to state variables in the model are also required for a testing of the whole model.

It may here be pointed out in parenthesis that it falls outside the scope of this work to give a complete discussion of most of the aspects that may enter the question in connection with the design of surveys for all the indicated problems.

In the following sections attention is drawn to some selected particular problems. In the next section (7.2) will be given a consideration of a problem connected with exploratory investigations. After this methods needed for the estimation of model parameters will occupy the focus of interest in section 7.3. In the concluding section 7.4 is given a brief discussion of the general problem connected with surveys, which includes the task of analysing the error in observations and calculated statistics with the existent needs as point of departure.

7.2. Exploration in econometrics[1]

7.2.1. *On the problem of drawing causal inferences from observational data*

In the previous section exploration in the real system was described as a method whereby to attempt on an empirical basis to get a foundation for the specification of a model. It is thus a general problem, which is tackled by trying to draw causal inferences from available observational data. Naturally, such a posing of the problem, aiming at insights in the causal nexus, provides an incentive to broad discussions in scientific fields and in various philosophical, statistical and other disciplines. By way of introduc-

[1] This section and, in particular, section 7.2.3. are based on Norlén (1971, 1972).

tion we shall also be touching upon some points in the discussion of the relevant partly or wholly controversial points of departure in this approach.

In the first place it must be regarded as perfectly natural that the attempt to get results in causal terms from statistical analysis should constitute a deep problem. The question implies, amongst other things, taking up an attitude to the causality concept. Causality was discussed in the 20's. A pessimistic standpoint in the debate was taken up by, inter alios, Russell (1914), who asserted that the causality concept ought to have no place in science. This standpoint and similar attitudes probably constitute the reason why the discussion of causality seems to have stagnated from the 20's until about twenty years ago.

Another and closely related question concerns the place of the specification problem in modelling. One argument against the attempt at an approach here referred to is that the problem of constructing a model is a problem on the theoretical or subject-matter plane and has nothing to do with the empirical evidence that may occur. There does not seem to be anything to object to this argument *per se*; cf. the discussion is section 2.3. The question of what knowledge a specification of the model is, or, should be, based upon does, however, arise. But—if only one of the bases or foundations in the process of building the "final" model may be considered to consist of observation of that which is modelled—this question is herewith eliminated. The number of requisite and preparatory empirical analyses will evidently depend on the amount of other knowledge given in advance. The actual empirical foundation in the search for a model is thus in many ways a matter of degree; if there is only little existing or available theoretical knowledge, then a "low-information approach" in the sense used by Wold (1968*a*) is the only way out at the first stages of model building.

Even from some formal statistical points of departure arguments have been adduced against the possibility of causal analysis of the kind here referred to. Arguments to the effect that cause–effect hypotheses are asymmetrical, whereas correlations and diagrams are symmetrical fall into this category. These assertions appear to be valid for individual cause–effect pairs. If, on the other hand, the number of variables is increased, then in some respect we get asymmetries, which may therewith be used in the analysis. An early attempt at using asymmetries when there are more than two variables dates back to Working (1933).

The above remarks refer chiefly to the field of the social sciences. In the natural sciences the discussion seems to have developed rather unaffected by developments in the social sciences. Wright (1934), for example, has indicated a method for the calculation of so-called path coefficients between variables which may be said to be causally inspired.

Russell's anti-causal impact in the social sciences seems to have been broken in the mid 50's. Early writers in this field are Simon (1953) and Wold (1952, 1954). Since then the question of causality appears in addition to have been given growing attention. Mention may be made of in various respects important works by Blalock (1961) and Bunge (1959). Contributions which have come to the forefront today come from the above-mentioned work by Wright (1934); see e.g. Tukey (1954), Duncan (1966), the contributions in Borgatta & Bohrnstedt (1969, 1970), Costner (1971) and Blalock (1971) to note only a few of the many works in this series. Statistical works dealing with this question have been referred to in some detail by the above-mentioned writers Blalock and Wold, and also by Eklund (1959).

Although critical comments could be made on the subject, some exploratory techniques seem to be worth applying. Apart from the constant need of getting "new ideas", the reason for this is twofold: the technical potentialities of the computers and a wealth of available data. This calls for a reevaluation of some techniques, e.g. the technique suggested by Wold (1961) and developed by Lyttkens (1964).

7.2.2. *Problem situation and aim of the study*

In the subsequent part of this section we shall discuss a particular aspect in connection with attempts to draw causal inferences from the statistical analysis of a set of observations. In this context causality will be based on a relatively broad view of the concept. On the whole, if the mode of expression is simplified, the problem will be formulated as an econometric attempt to obtain linear analytic relations between variables (quantities, criteria, measures) showing good agreement in the sense that the residuals are small.

The present problem of drawing causal inferences is thus a matter related to "the causal interpretation of econometric systems", which is a theme of old standing in econometrics.[1] Of the account given in this work two relevant links to causal interpretation are the following: (i) the logical structure of econometric systems and the structure of the present model systems (section 4.2.3) and (ii) simulation with the present model systems and fictitious experiments in modelled systems (section 4.3).

Let us consider the case in which the only information that is given consists of observations of the following two categories of variables:

$\boldsymbol{\eta}$ $n \times 1$ vector of endogenous variables (7.2.1)

$\boldsymbol{\zeta}$ $m \times 1$ vector of predetermined variables.

Let us assume that for these variables there is an econometric system of

[1] Meissner (1971) has reviewed this discussion in econometrics. Among other things, he has made clear the different aspiration levels employed: on the one hand, the low aspiration level, when only joint probability laws are specified, and on the other hand, the high aspiration level, when the statistical formulation is supplemented by causal interpretations of the variables and relations involved.

relations with a structural form (4.2.17) and a reduced form (4.2.18). It is further assumed that the classical specification of disturbances (4.2.23–24) is applicable. An assumption comparable with this is the predictor specification

$$E(\boldsymbol{\eta}\,|\,\boldsymbol{\zeta}) = \boldsymbol{\Omega}\boldsymbol{\zeta}. \tag{7.2.2}$$

This information is sufficient to enable us to estimate the parameters in the reduced form.[1] Thus the method of least squares gives consistent estimates of $\boldsymbol{\Omega}$ in (7.2.2).[2] In compact matrix form the estimator can with easily understood designations be written

$$\mathbf{W} = (\Sigma\mathbf{y}\mathbf{z}')(\Sigma\mathbf{z}\mathbf{z}')^{-1}. \tag{7.2.3}$$

The aim may now be formulated as an attempt to get a suitable structural form. This gives a basis for increasing the resolution level for the view of the system by one unit. Connecting up to the interpretation of the econometric system in section 4.2.3, the structural form is usable in connection with the specification of the system at the resolution level *one*, while the reduced form is compared with systems at the resolution level *zero*.

As point of departure for this attempt the estimate $\mathbf{W}$ from (7.2.3) is used. It is well-known that a structural and the reduced form for a system are not in 1–1 correspondence with each other. A unique structural form does not belong to a given reduced form. Let us consider the reduced form (4.2.18). If there is no further information, for example in the form of restrictions imposed on the coefficients in the structural form, then the reduced form defines the following class of structural forms

$$\mathbf{H}\boldsymbol{\eta} = \mathbf{H}\boldsymbol{\Omega}\boldsymbol{\zeta} + \mathbf{H}\boldsymbol{\epsilon}, \tag{7.2.4}$$

where $\mathbf{H}$ is a non-singular matrix, since (7.2.4) gives

$$\boldsymbol{\eta} = \mathbf{H}^{-1}(\mathbf{H}\boldsymbol{\Omega}\boldsymbol{\zeta} + \mathbf{H}\boldsymbol{\epsilon}) = \boldsymbol{\Omega}\boldsymbol{\zeta} + \boldsymbol{\epsilon}. \tag{7.2.5}$$

This, in econometrics, is generally referred to as the identification problem: "The identification problem may be stated as that of deducing the values of the parameters of the structural relations from a knowledge of the reduced-form parameters" (Johnston 1963, p. 240).

The following attempt to obtain a structural form from the estimate $\mathbf{W}$ of the parameters of the reduced form rests on a criterion of simplicity for the setting up of parameters in the structural form. The system is thus

---

[1] We are considering the case when there are no measurement errors (no "errors in variables"). Reference is made to Norlén (1972) for a discussion of the case when there are measurement errors.

[2] For a simple proof of consistency, see Wold (1963*a*).

presumed to be overidentified, i.e. there are fewer than or equal to m variables in every relation to the structural form. Furthermore, at least one of these variables must be an endogenous variable. In other words, this may be expressed as if there were at least $(m+n)-m=n$ unspecified zeros in every row in the coefficient matrix $[\mathbf{B} \vdots \mathbf{\Gamma}]$. This will be referred to as the simple-structure hypothesis. As an index of simplicity we use the number of zero parameters in the structural form. Thus, provided that every relation consists of at least two variables, the simplicity index will lie between n^2 and $n(n+m-2)$.[1]

The problem is therefore to try to find which variables form the structural relations, or, what amounts to the same thing, in which places in the coefficient matrices there are zero parameters. In econometrics, on analogy with the situation in factor analysis most closely correspoding to it, this is referred to as the rotation problem in econometrics.

The problem may be formulated as follows. Find two matrices $\mathbf{B}$ and $\mathbf{G}$ so that the distance between $\hat{\mathbf{y}}^*$ and $\mathbf{y}^*$

$$d(\hat{\mathbf{y}}^*, \mathbf{y}^*), \tag{7.2.6}$$

where

$$\begin{cases} \hat{\mathbf{y}}^* = \mathbf{B}^{-1}\mathbf{G}\mathbf{z} \\ \mathbf{y}^* = \mathbf{W}\mathbf{z} \end{cases} \tag{7.2.7}$$

is small according to some defined measure of distance. It has earlier been pointed out that with more than m variables in every relation there is a whole class of $\mathbf{B}$ and $\mathbf{G}$ matrices for which

$$\mathbf{B}^{-1}\mathbf{G} = \mathbf{W}. \tag{7.2.8}$$

Thus the distance (7.2.6) for these structural forms is equal to zero. For simple structures, however, the distance will generally be different from zero if one does not proceed from population quantities but calculates the distance with a sample of observations of the variables as point of departure.

7.2.3. *A direct selection technique*

For a given value of the index of simplicity let a procedure be envisaged that gives the structure which has the least distance. If the simplicity index is allowed to vary a set of structural forms will be obtained. These

[1] The only identifying restrictions to be discussed are thus the exclusion of certain variables from certain relations. For a treatment of the identification problem in econometrics, see e.g. Fisher (1966). The simple structure hypothesis made above is a slight reformulation of what is known as the order condition for (over)identifiability in econometrics. It should be remembered that the condition is, however, not sufficient, since the system is (over)identified only if $\mathbf{B}$ and $\mathbf{\Gamma}$ have a unique (an overdetermined) solution from $\mathbf{B}^{-1}\mathbf{\Gamma} = \mathbf{\Omega}$, that is, sufficiency also depends on the parameter values of the reduced form. Thus, if the system is not known, identifiability can not be assessed beforehand; see e.g. Johnston (1963).

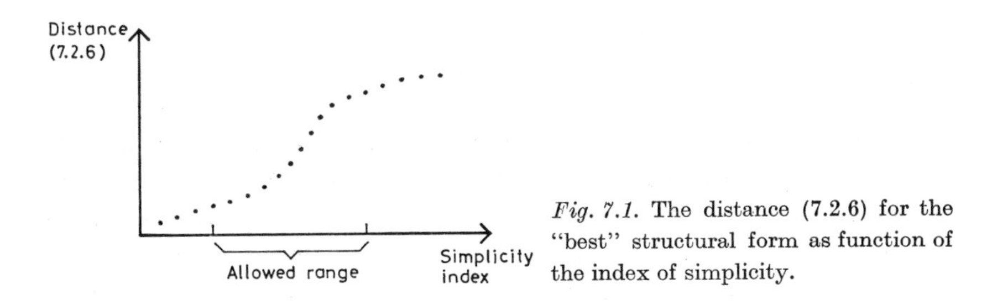

Fig. 7.1. The distance (7.2.6) for the "best" structural form as function of the index of simplicity.

forms will have monotonously increasing distance with increasing simplicity. The argument is illustrated in Fig. 7.1. For the moment we postpone the problem of determining the value of the index of simplicity.

For small systems it is possible to use a direct selection technique and minimize a measure (7.2.6) for all possible simple structures. The number of simple structures for one relation is

$$\sum_{k=2}^{m} \sum_{i=1}^{\min(k,n)} \binom{n}{i} \binom{m}{k-i}. \tag{7.2.9}$$

In Table 7.1 is given the value of (7.2.9) for some small systems. Among these one-relation structures n structures are thus to be chosen which are composed in such a way that the appurtenant estimate of the matrix B is not singular.

The distance (7.2.6) is defined as an expression in two vectors. These vectors represent two points in the n-dimensional Euclidian space. The squared distance between the vectors is

$$(\hat{\mathbf{y}}^* - \mathbf{y}^*)'(\hat{\mathbf{y}}^* - \mathbf{y}^*) = \mathrm{tr}\,\{(\hat{\mathbf{y}}^* - \mathbf{y}^*)(\hat{\mathbf{y}}^* - \mathbf{y}^*)'\}. \tag{7.2.10}$$

Summing over the observations, we get from (7.2.7) the following interpretation of the distance

$$d^2(\hat{\mathbf{y}}^*, \mathbf{y}^*) = \mathrm{tr}\,\{(\mathbf{B}^{-1}\mathbf{G} - \mathbf{W})\mathbf{M}(\mathbf{B}^{-1}\mathbf{G} - \mathbf{W})'\}, \tag{7.2.11}$$

Table 7.1. *Number of possible structures for one relation in some small relation-systems*

	Number of one-relation structures for m equal to						
n	2	3	4	5	6	7	8
2	5	16	39	86	181	372	755
3	9	31	80	184	399	837	1 722
4	14	52	143	346	780	1 684	3 537
5	20	80	235	601	1 417	3 169	6 838
6	27	116	364	986	2 440	5 678	12 649
7	35	161	539	1 547	4 025	9 773	22 556
8	44	216	770	2 340	6 404	16 248	38 939

where $\mathbf{M}$ is the symmetrical matrix

$$\mathbf{M} = \Sigma \mathbf{z}\mathbf{z}'. \tag{7.2.12}$$

The use of (7.2.11) as a measure of the distance, however, creates difficult non-linear problems. Minimum of (7.2.11) in terms of the unknown matrices $\mathbf{B}$ and $\mathbf{G}$ cannot be given explicitly, and numerical computations involve laborious calculations. This approach has therefore not yet been more carefully investigated. A simplified procedure for a solution is elaborated in the following way.

Consider the term in brackets in the right side of (7.2.11). After rearrangement, pre- and post-multiplication with $\mathbf{B}$ and $\mathbf{B}'$ respectively gives the following simple matrix

$$(\mathbf{BW} - \mathbf{G})\mathbf{M}(\mathbf{BW} - \mathbf{G})'. \tag{7.2.13}$$

If the trace of the above matrix is used as a measure of the distance every relation can be treated separately. In symbols

$$d(\hat{\mathbf{y}}^*, \mathbf{y}^*) = \sum_{i=1}^{n} (\mathbf{b}_i'\mathbf{W} - \mathbf{g}_i')\mathbf{M}(\mathbf{b}_i'\mathbf{W} - \mathbf{g}_i')', \tag{7.2.14}$$

where $\mathbf{b}_i$ and $\mathbf{g}_i$ are the parameter vectors in the ith relation.

Consider one relation-structure and introduce the following designations

$$\begin{cases} \mathbf{x} \text{ the vector of non-zero elements in } \begin{pmatrix} \mathbf{b} \\ \mathbf{g} \end{pmatrix} \\ \mathbf{C} \text{ corresponding symmetric submatrix of } \begin{pmatrix} \mathbf{W} \\ -\mathbf{I} \end{pmatrix} \mathbf{M} \begin{pmatrix} \mathbf{W} \\ -\mathbf{I} \end{pmatrix}'. \end{cases} \tag{7.2.15}$$

A term in the right side in the relation (7.2.14) can therewith be written

$$\mathbf{x}'\mathbf{C}\mathbf{x}. \tag{7.2.16}$$

This function is minimized, taking into account the normalization condition

$$\mathbf{x}'\mathbf{x} = 1. \tag{7.2.17}$$

Thus continue and consider the Lagrange function

$$f = \mathbf{x}'\mathbf{C}\mathbf{x} - l(\mathbf{x}'\mathbf{x} - 1), \tag{7.2.18}$$

where l is the Lagrange multiplier. The partial derivatives are given from

$$\frac{1}{2}\frac{\partial f}{\partial \mathbf{x}} = \mathbf{C}\mathbf{x} - l\mathbf{x}. \tag{7.2.19}$$

Put these derivatives as equal to zero. The solutions to the resulting equations are the characteristic vectors of the matrix $\mathbf{C}$. Since the value (7.2.16) is equal to

$$\mathbf{x}'\mathbf{C}\mathbf{x} = \mathbf{x}'l\mathbf{x} = l \tag{7.2.20}$$

that vector, accordingly, which belongs to the least characteristic root should be chosen.

From (7.2.20) we also get the following simple expression for the distance (7.2.14)

$$d(\hat{\mathbf{y}}^*, \mathbf{y}^*) = \sum_{i=1}^{n} l_i, \tag{7.2.21}$$

where l_i is the least characteristic root belonging to the structure of relation i.

To summarize, the technique consists in a calculation of the distance (7.2.21) for all possible structures. The structure which gives the least distance is then chosen. Further, the solution is found to be based on the theory for characteristic roots and vectors. In point of fact, the technique consists in seeking the least principal-component variance for every possible combination of $\mathbf{y}^*$- and $\mathbf{z}$-variables.[1] This can be seen from (7.2.15). In easily understood symbols

$$\begin{pmatrix} \mathbf{W} \\ -\mathbf{I} \end{pmatrix} \mathbf{M} \begin{pmatrix} \mathbf{W} \\ -\mathbf{I} \end{pmatrix}' = \begin{bmatrix} \mathbf{M}_{y^*y^*} & -\mathbf{M}_{y^*z} \\ -\mathbf{M}_{zy^*} & \mathbf{M}_{zz} \end{bmatrix} \tag{7.2.22}$$

Example 7.1. The selection technique we have described for the obtaining of structural form for small econometric systems has been investigated with the Monte Carlo method. Observations have been generated from a given structure, after which the technique has been applied. The procedure has been repeated a number of times. Each time one observes whether the structure sought has been obtained from the technique. The applicability of the technique is then measured by the frequency of correct solutions.

The frequency of correct solutions will of course depend upon the characteristics of the system and the amount of information obtained about the system, i.e. the measuring technique, as well as upon the properties of the actual technique applied.

A reasonable requirement, however, as regards the technique, is that the frequency should in any event on an average be greater than the frequency that is obtained if a random choice is made among the possible structures. Further, the technique should be of such a nature that the frequency of successful interpretations increases with increasing size of sample or, which amounts to the same thing, diminishing size of the residuals in the system. With the intention of ascertaining whether these statements are reasonable for the technique indicated, experiments have been carried out with the system

$$\begin{cases} \eta_1 - 0.8\eta_2 = 1.2 + 1.2\zeta_1 + 0.9\zeta_2 + \delta_1 \\ -0.75\eta_1 + \eta_2 = 1.5 + 0.5\zeta_3 + 1.1\zeta_4 + \delta_2 \end{cases} \tag{7.2.23}$$

---

[1] For a definition of principal components, see e.g. Anderson (1958).

Table 7.2. *Observations of the predetermined variables in the system (7.2.23) used in the Monte Carlo study*

Variable	Observation number							
	1	2	3	4	5	6	7	8
ζ_1	1	1	1	1	−1	−1	−1	−1
ζ_2	1	1	−1	−1	1	1	−1	−1
ζ_3	1	−1	1	−1	1	−1	1	−1
ζ_4	−1	−1	1	1	1	1	−1	−1

with two endogenous and four predetermined variables and constant terms. It further emerges that every systematic part in the structural form has four variables and one constant term. Let us assume that this information is available. In this case there will be

$$\sum_{i=1}^{2} \binom{2}{i} \binom{4}{4-i} = 14 \tag{7.2.24}$$

structures for every relation to choose between; cf. (7.2.9). From these 14 one-relation structures it is possible to choose one system with two relations in 39 different ways, so that the resulting system generally gives a non-singular estimate of the coefficient matrix B for the endogenous variables. In accordance with the discussion carried on earlier the frequency for successful use of the technique should thus on an average exceed 1/39.

The following experimental design was set up. Number of:

residual-sizes	4
replications/residual-size	100
observations/replication	8.

In every situation the eight observations of the predetermined variables are given according to Table 7.2.

The residuals are generated from the bivariate normal distribution with the mean values equal to zero and dispersion matrix $\sigma^2 \cdot \mathbf{I}$. The four values for σ^2 are so chosen that the predetermined variables "explain" the endogenous variables with 25, 50, 75 and 100 per cent accuracy. As a measure of the explanation we use in this connection the quotient of the determinants

$$\frac{\left|\begin{matrix}\text{dispersion matrix for the expected values}\\ \text{of the endogenous variables}\end{matrix}\right|}{|\text{dispersion matrix for the endogenous variables}|}. \tag{7.2.25}$$

From Table 7.2 it follows that the dispersion matrix for the predetermined variables in this case is the identity matrix. The four residual variances σ^2 are therewith easily obtained from (4.2.18) and (7.2.2) as the solutions of the equation

$$\frac{|\mathbf{\Omega\Omega}'|}{|\mathbf{\Omega\Omega}'+\sigma^2\cdot\mathbf{I}|}=R, \tag{7.2.26}$$

where

$$R=0.25,\ 0.50,\ 0.75 \text{ and } 1.00. \tag{7.2.27}$$

Since there are constant terms in the system we are operating on normalized variables with the mean values put equal to zero.

In Fig. 7.2 below will be found a summary of the results of the investigation.[1] The frequencies obtained—f—may be seen as an estimate of the probability P in the relation

$$f=P+\varepsilon_P, \tag{7.2.28}$$

where P is the probability of making a correct choice of the structure in the system. The probability is in general a function of characteristics of the system, the measuring instruments used and the technique applied. The residual ε_P is the error in the estimation of P. In the Figure the estimate of P is given as a function of the degree of explanation R, defined by (7.2.26), which is a characteristic of the system. The results show that the frequencies obtained are considerably higher than the frequencies that would be expected if random choices were made among the possible structures. The results also show that the frequencies increase when the residuals in the system are smaller. In the case in which R is equal to one, i.e. where all residuals are equal to zero and the system is a deterministic system, the technique seeks out the right structure for each use. It would be of value to extend the above illustration to a more comprehensive and complete study which should take into account several arguments in the probability function P in (7.2.28). A realistic problem of topical interest in this connection would be that entailed in a more detailed investigation of the consequences of the measuring instruments used in the event that observations obtained show measurement errors. We shall revert to this question in section 7.4.

7.2.4. *Determination of the value of the index of simplicity*

Up to now we have not considered the problem of determining the value of the index of simplicity. One way is to see this as a testing problem and derive the distribution of d in (7.2.6). For the suggested selection technique this implies the derivation of the distribution of the sum of characteristic

[1] The computations in this example and in the following examples 7.2 and 7.4 were performed on the university computer CDC 3600 in Uppsala.

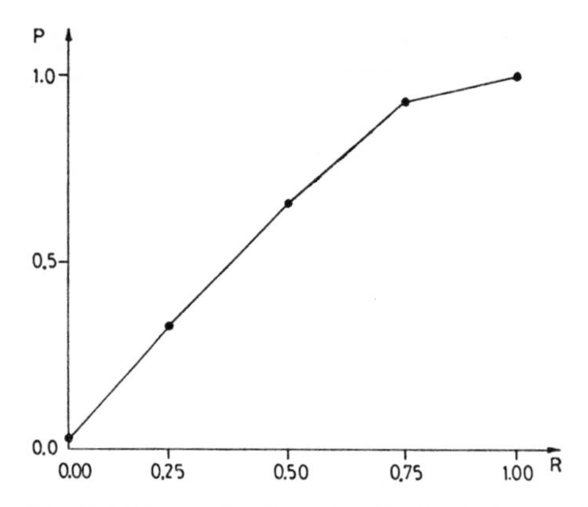

Fig. 7.2. The results from the Monte Carlo experiment with the selection technique indicated and used with the system (7.2.23). The Figure shows the estimates of the probability P of interpreting the structure for the system correctly, as a function of the degree of explanation R in the system defined by (7.2.26) for the number of observations equal to eight.

roots in (7.2.21). We shall not treat this problem in detail. For large samples, however, the following observation makes up a partial solution to this problem.

Consider first the matrix (7.2.22). Assume that the predetermined variables have zero means and dispersion matrix $\mathbf{\Psi}$. The expectation of (7.2.22) is

$$\mathbf{\Phi} = \begin{bmatrix} N\mathbf{\Omega}\mathbf{\Psi}\mathbf{\Omega}' + m\mathbf{\Sigma} & -N\mathbf{\Omega}\mathbf{\Psi} \\ -N\mathbf{\Psi}\mathbf{\Omega}' & N\mathbf{\Psi} \end{bmatrix}, \tag{7.2.29}$$

where $\mathbf{\Omega}$ is the matrix consisting of reduced form coefficients, N is the number of observations, m is the number of predetermined variables and $\mathbf{\Sigma}$ is the dispersion matrix of the disturbances in the reduced form.

In the next place consider a structure for one relation in the system. Assume that it is correctly specified and apply the minimization (7.2.16–20) with the population matrix (7.2.29) used instead of (7.2.22). For the least characteristic root in the population the following inequality is valid

$$\lambda \leqslant \begin{pmatrix} \boldsymbol{\beta} \\ \boldsymbol{\gamma} \end{pmatrix}' \mathbf{\Phi} \begin{pmatrix} \boldsymbol{\beta} \\ \boldsymbol{\gamma} \end{pmatrix}. \tag{7.2.30}$$

If we make use of (7.2.29) and of the identity

$$\boldsymbol{\beta}'\mathbf{\Omega} = \boldsymbol{\beta}'\mathrm{B}^{-1}\mathbf{\Gamma} = \boldsymbol{\gamma}' \tag{7.2.31}$$

the inequality (7.2.30) is reduced to

$$\lambda \leqslant m\boldsymbol{\beta}'\mathbf{\Sigma}\boldsymbol{\beta}. \tag{7.2.32}$$

If (i) the sample matrix (7.2.22) is interpreted as the dispersion matrix from N observations on a random variable distributed $\mathrm{N}(\mathbf{0},\mathbf{\Phi})$ and (ii) the considered submatrix of $\mathbf{\Phi}$ has distinct characteristic roots, the least characteristic root l for N large is distributed $\mathrm{N}(\lambda, 2\lambda^2/N)$; see e.g. Morrison (1967). Hence an upper bound of l can be obtained from

$$l \leqslant (1 + \sqrt{2}\, x_p/\sqrt{N})\, m\mathbf{b}'\mathbf{S}\mathbf{b}, \tag{7.2.33}$$

where x_p denotes the upper $100\,p$ percentage point of the standard normal distribution, $\mathbf{b}$ consists of zeros and elements from the solution vector $\mathbf{x}$ and where $\mathbf{S}$ is an estimate of the dispersion matrix $\mathbf{\Sigma}$.

For the correct structure the value of l is consequently of the order 1. For incorrectly specified relation structures one may further expect that l is of the order N, where N is the number of observations. We can use this result and exclude those not probable relations for which the inequality (7.2.33) is not valid. Since the distance (7.2.21) is monotonously increasing with increasing values of the index of simplicity, it also follows that for some of the considered values of the simplicity index there may not exist an acceptable system. The argument is made clear by the application presented below.

Example 7.2. The selection technique has been applied to sociological cross-section data. The data are taken from a study by Gelin (1969, 1972) dealing with problems centered on female working situations in Uppsala. The population is defined as those women in the age interval 20–59 years who lived in the town of Uppsala and some of its surrounding communes in October 1967. A random sample of 1 064 persons was drawn from the Uppsala County General Insurance Register, and 906 interviews passing a quality control were performed, which corresponds to 85 per cent of the whole sample. The variables in the example to be presented below are defined as follows:
Three endogenous variables:

y_1 = *Income after tax* (Swedish kronor). For those persons in the sample who do not earn money outside the family this variable has the value 0.

y_2 = *general attitude score,* obtained from a composite attitude-scale designed to measure the general attitude towards the division of labour between the sexes. This was built in a multi-step way: 35 items, the majority of which had been included in other studies in Scandinavia, were factor-analysed and put into Guttman-scaling procedures resulting in five separate scales. A person's score for this variable is the sum of her scores on each of three of the five scales; the higher the value, the more positive was the attitude towards women working, desegregation between the sexes, and a more equal distribution of power between men and women.

y_3 = *Number of hours worked per week.* Analogously to y_1, this variable takes the value 0 for those who do not perform paid work outside the family.

Four predetermined variables:

z_1 = *Achieved occupational training score.* This variable was coded in the following way:

$$z_1 = \begin{cases} 1 & \text{no occupational training} \\ 2 & \text{some occupational training, but for less than one year} \\ 3 & \text{between one and two years of occupational training} \\ 4 & \text{more than two years of occupational training} \\ 5 & \text{more than five years of occupational experience in an occupation} \end{cases}$$

z_2 = *Achieved formal education score.* This variable was coded in the following way:

$$z_2 = \begin{cases} 1 & \text{primary school} & \text{("folkskola")} \\ 2 & \text{higher primary school} & \text{("högre folkskola")} \\ 3 & \text{middle school} & \text{("realskola")} \\ 5 & \text{matriculated} & \text{("studentexamen")} \\ 7 & \text{university degree} & \text{("akademisk examen")} \end{cases}$$

z_3 = *Number of children in the age-group 0–16 years.*

z_4 = *Reference group attitude index score.* This variable is a composite index consisting of the sum of two separate measures: the person's perception of (i) her mother's attitude towards married women working and (ii) her female friends' attitudes towards married women working. A higher score for this index corresponds to a more negative attitude towards married women working.

A discussion of the metrical properties of the variables cannot be elaborated here, since it requires, among other things, a presentation of the theoretical concepts involved; see Gelin (1972). As a short-cut judgement we might make the assertion that the variables y_1, y_3 and z_3 have interval-scale properties at least "on the surface", while the variables y_2 and z_2 are of the approximate interval-scale type and the variables z_1 and z_4 could be located somewhere on the borderline between interval and ordinal scales.

The analysis presented below is performed on a sub-category of the sample and is part of the testing of a set of sociological hypotheses concerning variations in the strength of connections between ideological beliefs and goals on the one hand, and actual behaviour on the other. The sub-category in the example consists of women in the age-group 20–39 years, not married, not studying and having a specified score on a separate attitude-scale constructed to throw light on the set of hypotheses indicated above concerning the relations between ideology and actual behaviour.

We operate on normalized variables with means $=0$ and variances $=1$. The following correlation matrix, calculated from 55 observations belonging to the sub-category mentioned, was obtained.

$$\begin{array}{c} \\ y_1 \\ y_2 \\ y_3 \\ z_1 \\ z_2 \\ z_3 \\ z_4 \end{array} \begin{array}{ccc:cccc} y_1 & y_2 & y_3 & z_1 & z_2 & z_3 & z_4 \\ \hline 1.000 & & & & & & \\ 0.369 & 1.000 & & & & & \\ 0.784 & 0.218 & 1.000 & & & & \\ \hdashline 0.270 & 0.316 & 0.185 & 1.000 & & & \\ 0.617 & 0.487 & 0.262 & -0.052 & 1.000 & & \\ 0.019 & -0.010 & -0.164 & 0.086 & 0.014 & 1.000 & \\ -0.016 & 0.042 & 0.076 & 0.268 & -0.152 & 0.028 & 1.000 \end{array} \tag{7.2.34}$$

In accordance with the basic assumptions behind the selection technique, it is assumed that there exists an econometric three-relation system with a structural form (4.2.17) and a reduced form (4.2.18) for the described variables for which the predictor assumption (7.2.2) applies. The method of least squares gives the following estimates of the parameters in the reduced form

$$\begin{pmatrix} y_1 \\ y_2 \\ y_3 \end{pmatrix} = \begin{bmatrix} 0.30 & 0.63 & -0.02 & 0.00 \\ 0.34 & 0.51 & -0.05 & 0.03 \\ 0.20 & 0.29 & -0.19 & 0.07 \end{bmatrix} \begin{pmatrix} z_1 \\ z_2 \\ z_3 \\ z_4 \end{pmatrix} + \begin{pmatrix} e_1 \\ e_2 \\ e_3 \end{pmatrix}. \tag{7.2.35}$$

As measured by the multiple correlation coefficient the "degrees of explanation" in the three relations were obtained as 0.69, 0.60 and 0.38 respectively.

The selection technique is now to be applied. The number of possible structures for one relation in the structural form is according to (7.2.9) for three endogenous and four predetermined variables:

$$\sum_{k=2}^{4} \sum_{i=1}^{\min(k,3)} \binom{3}{i} \binom{4}{k-i} = 80. \tag{7.2.36}$$

If the rule associated with the result (7.2.34) is applied, where we have put $x_p = 0$, the number of possible one-relation structures is reduced to 56. In Table 7.3 is given the distribution of the possible three-relation system structures for these 56 one-relation structures.

From the Table we conclude that the number of possible values of the index of simplicity has been reduced to five from the original six

Table 7.3. *The distribution of the number of possible three-relation systems for 56 selected one-relation structures from the 80 possible*

The number of possible systems consisting of all possible combinations from the original 80 one-relation structures is given in brackets

	Simplicity index (=number of zero parameters in the structural form)						
System	9	10	11	12	13	14	Total
VR-system	14 (60)	30 (252)	21 (480)	3 (474)	0 (228)	0 (36)	68 (1 530)
CC-system	1 382 (2 220)	2 191 (4 926)	1 344 (4 680)	310 (1 962)	12 (300)	— (—)	5 239 (14 088)
ID-system	1 776 (2 016)	1 816 (2 496)	640 (1 128)	66 (168)	— (—)	— (—)	4 298 (5 808)
Total	3 172 (4 296)	4 037 (7 674)	2 005 (6 288)	379 (2 604)	12 (528)	0 (36)	9 605 (21 426)

considered; the most "simple" structure is thus excluded. Application of the technique gives a "best" structural form in each of the five cases. The selected systems are shown in Fig. 7.3.

Since a test for the value of the index simplicity has not as yet been fully developed, a subjective choice among the five structures must be

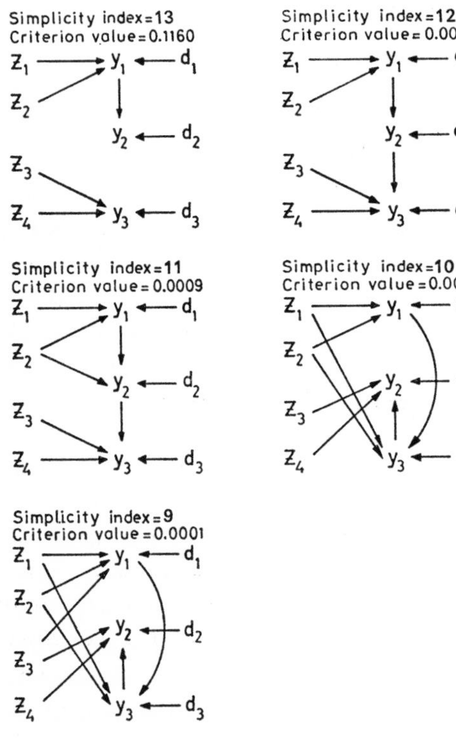

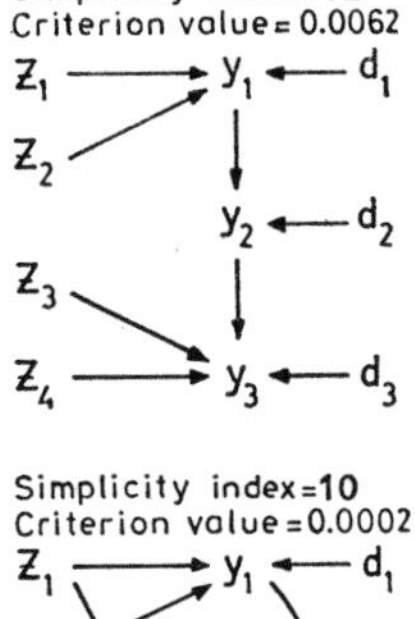

Fig. 7.3. The selected structural form for the five considered values of the index of simplicity. The criterion values are the values of (7.2.21) divided by the number of observations.

made. We chose the CC-system as obtained for the value 11 of the simplicity index partly in view of the fact that this structure gives a large percentual decrease in the criterion value when the criterion values for consecutive degrees of simplicity are compared.[1]

The method of least squares, applied relation for relation, gives the following parameter estimates for this structure:[2]

$$\begin{cases} y_1 = 0.30z_1 + 0.63z_2 + d_1 \\ y_2 = 0.11y_1 + 0.42z_2 + d_2 \\ y_3 = 0.21y_2 - 0.16z_3 + 0.08z_4 + d_3. \end{cases} \tag{7.2.37}$$

The multiple correlation coefficients were obtained as 0.69, 0.50 and 0.29 respectively. Inspection of the selected structural form, considering, among other things, the signs of the parameter estimates obtained, gives some credibility to the structure obtained.

One might also reason in terms of stability-recurrence, with all the selected systems for the different values of the index of simplicity as point of departure: if the outcomes are "shaky", that is, if a lot of the selected relations change abruptly from level to level (e.g. if y_1 and z_4 are in the first relation on one level and y_1, z_1 and z_3 on the next level and y_1, y_3 and z_2 on the following level and so on), we should assume that this reflects the lack of a genuine structure in the data. If, on the other hand, one gets a core of the same variables recurring in each relation one might guess that this indicates the presence of a genuine structure in the data. For more elaborated theoretical and empirical aspects of the particular case, I refer to a forthcoming report by Gelin (1972).

---

[1] The relationship between the problem of determining the number of factors in factor analysis and the problem of determining the "appropriate" value of the simplicity index in the selection technique presented here might be an interesting problem area to investigate. A "jump" in the sequence of criterion-values associated with consecutive values on the index of simplicity seems to occur quite often in the empirical data so far analyzed, at a point where the structure, as judged by non-formal criteria (theoretical/empirical more or less well-founded hypotheses about the correct structure), seems to "fit". This phenomenon might be compared with the so-called "discontinuity criterion" used as a guide for choosing the number of factors in factor analysis of which Rummel (1970, pp. 364–365) remarks: "To my knowledge, no mathematical proof exists for this notion ... Cattell ... has pointed out, however, and I have noticed in my own factor analyses, that this discontinuity usually does appear within the neighbourhood of the eigen-value one cutoff and that it seems to discriminate quite well between common factors, on the one hand and unique or substantively meaningless factors, on the other. Discontinuity appears, therefore, as a useful criterion to employ when there are several eigen-values close to unity ..." (personal communication with Gelin (1972); see also subsequent section 7.2.5).

[2] The rationale for the application of the method of least squares in the present case of an CC-system is based on Wold (1961, 1964).

7.2.5. *Reformulations of the problem*

In this section we reformulate the rotation problem in econometrics. The rotation problem was formulated above as a problem of minimization. A solution was then obtained with the help of a selection technique.

A first observation is that we could select the structures having the greatest likelihood. For a recent application of the maximum likelihood principle in econometrics, assuming normally distributed variables, see Jöreskog (1972). One advantage of this formulation is that a (large sample) likelihood ratio test for a system may be applied. This test may be used in connection with, inter alia, the determination of the value of the index of simplicity.

For big systems, however, direct selection—with the selection technique described in section 7.2.3 or with selection based on the above application of the maximum likelihood principle—is all too time-consuming and costly, even with the help of a high-speed computer. For this reason the following two trains of thought may be worth closer consideration. Every comment indicates a possible escape from the method of systematically going through all possible structures.

In the first place, it is adduced that the above formulation of the problem agrees in principle with that used in seeking for the "best" regression equation, i.e. in the special case in which one has only one endogenous variable. One of these techniques is the so-called stepwise regression procedure. A description of this technique is available in, inter alios, Efroymson (1962). The way is thus open for reasoning on the present problem from analogy, with these techniques as point of departure.

In the second place, another solution may be offered through the following reformulation of the problem: Estimate $\mathbf{B}$ in the matrix consisting of coefficients in the structural form $\mathbf{B}[\mathbf{I} \vdots \mathbf{W}] = \mathbf{B}\mathbf{W}^*$, so that a simple structure is obtained. If the relations are normalized with diag $\{\mathbf{B}\mathbf{B}'\} = \mathbf{I}$ it becomes possible to treat the relations separately as before. Write one row in $\mathbf{B}\mathbf{W}^*$ as $\mathbf{b}'\mathbf{W}^*$. The problem here implies the finding, for every relation, of a point on the sphere $\mathbf{b}'\mathbf{b} = 1$ in the n-dimensional space that is spanned by the unknown coefficients $\mathbf{b}$, so that many elements in $\mathbf{b}'\mathbf{W}^*$ are close to zero.

Formulated in this way, the problem resembles the rotation problem in factor analysis. If all variables are normalized with zero means and variances equal to one, then the elements in $\mathbf{W}$ are partial correlation coefficients. All elements in $\mathbf{W}^*$ are therewith less than or equal to one in absolute value. The condition for normalization $\mathbf{b}'\mathbf{b} = 1$ ensures that the parameters $\mathbf{b}'\mathbf{W}^*$ are less than or equal to $\sqrt{n}$ in absolute value. It is thus possible to develop methods resembling the rotation techniques in factor analysis.

7.2.6. *Extensions of the selection technique*

We have hitherto designed the selection technique for problem situations which may be assigned to the low *a priori* information part in the scale shown in Fig. 7.4.

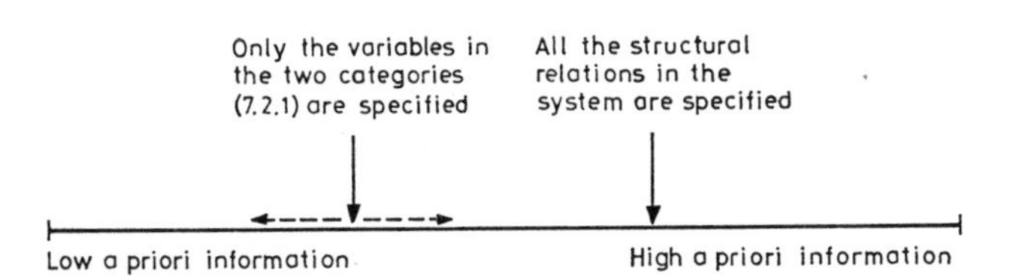

Fig. 7.4. Illustration of two levels of *a priori* information to be put into an econometric structure.

In further developments it is reasonable to provide for the use of more knowledge about the structure prior to the application of the method, e.g. in the form of wholly or partly specified structures for some of the relations. As for the selection technique described, such an extension can be made. The inclusion of additional information may be seen as restricting the search to a subset of the possible system structures.[1]

To summarize the results in this section, the foregoing is a presentation of a technique that may be used in attempts to get information concerning the real system. Where the technique is used successfully we thus get help, in the light of the discussion in section 4.2.3, in choosing (specifying) model mechanisms among the set of mechanisms that at an early stage in the modelling work are already supported by prior empirical evidence.

It should also be mentioned that the posing of the problem, and thus also the technique proposed, is of interest in connection with the reduction of an already existing and complex model to a simpler model. The task here is to try to make a selection of the most significant variables in the model and the probable relations between these variables. Thus, although it is presented in this chapter, the method may be applied also in the context of model analysis.

7.3. Estimation of parameters

7.3.1. *Background*

In a situation where the model is constructed and the goals for analysis are formulated the estimation of parameters appears as a clear-cut problem. If the procedures for estimation described in the literature are compared with the briefly discussed principles for model building we find that these are in marked contrast with each other. In the formulation of the model a structure for the latter is postulated, in which connection also the prerequisities for the postulated structure are given. The task of estimation therefore includes as a prime requirement a more detailed analysis of the consequences of the formulation.

[1] An extension to a still lower *a priori* information and a more exploratory approach, on the other hand, would be to allow for the reclassification of the variables (7.2.1) into new sets of endogenous and predetermined variables.

Among the points of departure for estimation we have also general principles—we are here referring chiefly to the methods of least squares (LS) and maximum likelihood (ML)—which have a great range in connection with the applications. Formerly, the numerical calculations in the estimation entailed much time-consuming work. Thanks to developments in the computer field this is no longer the case. The computer has now entered the arena as a supereminent aid and has for the most part relegated the work of calculation to the field of technical routine-work.

In view of what has been adduced in the foregoing we shall concentrate on a compilation of the existing needs for the estimation of parameters that emerged in the contexts of testing and analysis outlined, with the LS- and ML-methods for estimation.

7.3.2. *The methods of least squares and maximum likelihood*

In a general formulation of the problem of estimation we make the assumption that there are two sets of variables: endogenous variables $\boldsymbol{\eta}$ and predetermined variables $\boldsymbol{\zeta}$. For the endogenous variables it is assumed that a conditional frequency-function

$$f(\boldsymbol{\eta} \mid \boldsymbol{\zeta}, \boldsymbol{\alpha}) \tag{7.3.1}$$

exists, where $\boldsymbol{\alpha}$ is a parameter vector. The conditional expectation of $\boldsymbol{\eta}$ for a given $\boldsymbol{\zeta}$ is written

$$E(\boldsymbol{\eta} \mid \boldsymbol{\zeta}, \boldsymbol{\alpha}) = \mathbf{g}(\boldsymbol{\zeta}, \boldsymbol{\alpha}) \tag{7.3.2}$$

From observations of the variables we are to calculate an estimate $\mathbf{a}$ of the parameter $\boldsymbol{\alpha}$.

Taking either of the formulations (7.3.1) or (7.3.2) as point of departure, we can apply the LS- and ML-methods. An application of the LS-method proceeds in the main from (7.3.2) and consists in finding the $\mathbf{a}$ which minimizes a function of the elements in the main diagonal in the matrix

$$\Sigma(\mathbf{y}-\mathbf{g}(\mathbf{z}, \mathbf{a}))(\mathbf{y}-\mathbf{g}(\mathbf{z}, \mathbf{a}))', \tag{7.3.3}$$

where the summation is extended over the observations $\mathbf{y}$ and $\mathbf{z}$.

In the case of independent observations the ML-method consists in finding the $\mathbf{a}$ maximizing the likelihood function

$$L = \prod f(\mathbf{y} \mid \mathbf{z}, \mathbf{a}), \tag{7.3.4}$$

where the product is extended over the observations.

For clarifying insights into the prerequisites for and the properties of these methods standard statistical works by, inter alios, Cramér (1945) and Kendall & Stuart (1961) are recommended.

7.3.3. *Compilation with existing need for estimation*

A summarizing retrospect of the existing needs for estimation shows that the estimation of parameters in the model system, when the mechanisms are formulated in the form of conditional distributions as in (4.2.6), may be undertaken with the ML-method direct.

If, on the other hand, the mechanisms are specified in the form of conditional expectations—or if conditional expectations are deduced from the distributions—then the LS-method is applicable. It emerges further that the representation of system-behaviour in section 4.2.4 is compatible with the formulation (7.3.2). The same applies also for the analytic models in Chapter 6. In sections 6.2 and 7.2 simple illustrations are given of the use of the LS-method in the form of the two estimators (6.2.12) and (7.2.3).

Where the estimation refers to a calculation of the values for model parameters the predetermined variables will for the most part need to be treated as stochastic. In the treatment of analytic models as in Chapter 6 the predetermined variables are not stochastic, as such variables are given values of factors in simulation experiments.

A comparison between the LS- and ML-methods described reveals that the ML-method requires a relatively more solid basis in connection with the application, as the form for the frequency-function here needs to be specified, whereas the LS-method only requires specification of the expectation in the conditional distribution. It may thus be assumed that the LS-method is less sensitive with respect to erroneous specification. To this assumption may be added the practical point that LS-estimates are often simpler to calculate and test. These considerations motivate the use of the LS-method above all in the initial stage, when the mechanisms in the model are not completely known.

7.3.4. *Estimation in econometrics*

As we have seen, econometric structures are key instruments in the present approach. We make the following comments as regards the problem of estimating parameters in econometric relations.

Estimation of the classical econometric structural form (4.2.17) has been given comparatively much attention in the literature. One of the reasons for this is that the parameters in the structural form can in general not with advantage be estimated direct with the help of the LS-method. A well-known result by Haavelmo (1943) shows that the LS-method gives biassed estimates if it is applied relation by relation in an ID-system. This question has been discussed by Wold (1961), who shows that the structural form can in general not be formulated in terms of conditional expectations as in (7.3.2); cf. section 4.2.3. About ten methods have been devised to overcome the bias of the ordinary LS-method. We refer here to standard textbooks, e.g. those by Johnston (1963) or Goldberger (1964), for a review of the most commonly applied methods. There are specific advantages to recommend each of the methods referred to. However, one can make the same general assessment as the one given at the end of Chapter 6, i.e. with the assumption of independent observations all the methods yield estimates that have standard errors in inverse proportion to $\sqrt{N}$, where N is the number of observations. One may here adduce, by way of illustration, a method presented in section 7.2.3 which is on a par with this: the estimates obtained in the form of the characteristic vector corresponding to the least principal-component variance from the moment matrix of the calculated values for the expectation of the endogenous variables and the predetermined variables in every relation.[1]

As regards the generalized specification (4.2.28), with its fewer requisite assumptions concerning the distribution of the disturbances, a number of estimation techniques have been developed also for this formulation of the problem, which are based upon the LS-method and are characterized by iterative calculation procedures.[2]

In the following example LS- and ML-estimators of a VR-system are considered. We may here adduce some results from Norlén (1966). In addition, a simple proof that the standard errors of ML-estimates are less than those of LS-estimates is presented.

Example 7.3. Consider a VR-system with n relations. These relations may be written[3]

$$\underset{n\times 1}{\boldsymbol{\eta}} = \underset{n\times k}{\mathbf{Z}}\,\underset{k\times 1}{\boldsymbol{\alpha}} + \underset{n\times 1}{\boldsymbol{\epsilon}}, \tag{7.3.5}$$

where the elements in the matrix $\mathbf{Z}$ consist of predetermined variables and zeros. The disturbances are defined by

$$E(\boldsymbol{\eta}\,|\,\mathbf{Z}, \boldsymbol{\alpha}) = \mathbf{Z}\boldsymbol{\alpha}, \tag{7.3.6}$$

i.e. the system is given in the form (7.3.2). For the sake of simplicity it is assumed that the predetermined variables are not stochastic. Let us assume further that estimates are to be calculated from N independent observations

$$\{\mathbf{y}_i, \mathbf{Z}_i\}_{i=1}^{N}. \tag{7.3.7}$$

In the light of the prerequisite conditions we find that for these observations the following relations apply

$$\mathbf{y}_i = \mathbf{Z}_i\boldsymbol{\alpha} + \boldsymbol{\epsilon}_i \tag{7.3.8}$$

---

[1] This result, which may be called an extension of the so-called indirect least-squares method, is presented earlier in Norlén (1966).

[2] Wold (1965). For an investigation and extensions of this approach, see Mosbaek & Wold (1970) and Ågren (1972).

[3] With some small changes in the exposition, the following estimators may also be shown to give consistent estimates of parameters in CC-systems. See Wold (1961).

with

$$E(\mathbf{y}_i | \mathbf{Z}_i, \boldsymbol{\alpha}) = \mathbf{Z}_i \boldsymbol{\alpha}, \tag{7.3.9}$$

where $\boldsymbol{\epsilon}_i$ stands for the disturbances that are not observable.

In an application of (7.3.3) the following expression is set up for the calculation of LS-estimates

$$S = \sum_{i=1}^{N} (\mathbf{y}_i - \mathbf{Z}_i \mathbf{a})'(\mathbf{y}_i - \mathbf{Z}_i \mathbf{a}). \tag{7.3.10}$$

In clear cases indication of observations and sum and product signs will in the sequel be omitted. Derivation of (7.3.10) gives

$$\frac{1}{2}\frac{\partial S}{\partial \mathbf{a}} = -\Sigma \mathbf{Z}'\mathbf{y} + \Sigma \mathbf{Z}'\mathbf{Z}\mathbf{a}. \tag{7.3.11}$$

If the derivatives are put equal to zero one gets the LS-estimator

$$\mathbf{a}_{LS} = (\Sigma \mathbf{Z}'\mathbf{Z})^{-1} \Sigma \mathbf{Z}'\mathbf{y}. \tag{7.3.12}$$

The expectation of this estimator and the dispersion matrix are obtained in a simple way by combining (7.3.12) with (7.3.8)

$$\mathbf{a}_{LS} = (\Sigma \mathbf{Z}'\mathbf{Z})^{-1} \Sigma \mathbf{Z}'(\mathbf{Z}\boldsymbol{\alpha} + \boldsymbol{\epsilon}) = \boldsymbol{\alpha} + (\Sigma \mathbf{Z}'\mathbf{Z})^{-1} \Sigma \mathbf{Z}'\boldsymbol{\epsilon} \tag{7.3.13}$$

which combined with (7.3.9) gives

$$E(\mathbf{a}_{LS}) = \boldsymbol{\alpha} \tag{7.3.14}$$

and

$$D(\mathbf{a}_{LS}) = (\Sigma \mathbf{Z}'\mathbf{Z})^{-1} \sum_{i=1}^{N} \sum_{j=1}^{N} \mathbf{Z}_i' E(\boldsymbol{\epsilon}_i \boldsymbol{\epsilon}_j') \mathbf{Z}_j (\Sigma \mathbf{Z}'\mathbf{Z})^{-1}$$

$$= (\Sigma \mathbf{Z}'\mathbf{Z})^{-1} \Sigma \mathbf{Z}'\boldsymbol{\Omega}\mathbf{Z} (\Sigma \mathbf{Z}'\mathbf{Z})^{-1} \tag{7.3.15}$$

if the disturbances are homoscedastic with the dispersion matrix $\boldsymbol{\Omega}$.

With the intention of deriving the ML-estimator, let us assume that the residuals are normally distributed and homoscedastic with $\boldsymbol{\Omega}$ non-singular. The expression (7.3.1) may therewith be written

$$L = \prod \frac{1}{(2\pi)^{n/2} |\mathbf{W}|^{\frac{1}{2}}} \exp\{-\tfrac{1}{2}(\mathbf{y} - \mathbf{Z}\mathbf{a})' \mathbf{W}^{-1} (\mathbf{y} - \mathbf{Z}\mathbf{a})\}. \tag{7.3.16}$$

If the derivatives of the logarithm of the above function are put equal to zero, one gets the following non-linear likelihood equation system

$$\begin{cases} \mathbf{a}_{ML} = (\Sigma \mathbf{Z}'\mathbf{W}_{ML}^{-1}\mathbf{Z})^{-1} \Sigma \mathbf{Z}'\mathbf{W}_{ML}^{-1}\mathbf{y} \\ \mathbf{W}_{ML} = \dfrac{1}{N} \Sigma (\mathbf{y} - \mathbf{Z}\mathbf{a}_{ML})(\mathbf{y} - \mathbf{Z}\mathbf{a}_{ML})'. \end{cases} \tag{7.3.17}$$

A consequence of the non-linear features of this system is that it is in general not possible to write the solution in an explicit analytic form in the observations, as was the case for the LS-estimator (7.3.12). It can be shown, however, that with fairly general preconditions there does exist a solution which with every accuracy prescribed in advance can be obtained in a finite number of steps from the following iterative procedure

$$\begin{cases} \mathbf{a}_{ML}(s+1) = (\Sigma \mathbf{Z}' \mathbf{W}_{ML}^{-1}(s) \mathbf{Z})^{-1} \Sigma \mathbf{Z}' \mathbf{W}_{ML}^{-1}(s) \mathbf{y} \\ \mathbf{W}_{ML}(s) = \frac{1}{N} \Sigma (\mathbf{y} - \mathbf{Z}\mathbf{a}_{ML}(s)) (\mathbf{y} - \mathbf{Z}\mathbf{a}_{ML}(s))', \end{cases} \tag{7.3.18}$$

which may start from arbitrarily chosen starting values $\mathbf{a}_{ML}(0)$, where $\mathbf{a}_{ML}(s)$ and $\mathbf{W}_{ML}(s)$ indicate the values at iteration step number s.[1, 2]

Let us for the sake of simplicity assume that the dispersion matrix $\mathbf{\Omega}$ of the disturbances is known. The ML-estimator will be

$$\mathbf{a}_A = (\Sigma \mathbf{Z}' \mathbf{\Omega}^{-1} \mathbf{Z})^{-1} \Sigma \mathbf{Z}' \mathbf{\Omega}^{-1} \mathbf{y}, \tag{7.3.19}$$

which is also the so-called Aitken estimator (Aitken 1934). In the same way as for the LS-estimator the Aitken estimator can be shown to be unbiassed

$$E(\mathbf{a}_A) = E[\Sigma(\mathbf{Z}' \mathbf{\Omega}^{-1} \mathbf{Z})^{-1} \Sigma \mathbf{Z}' \mathbf{\Omega}^{-1} (\mathbf{Z}\boldsymbol{\alpha} + \boldsymbol{\epsilon})] = \boldsymbol{\alpha} \tag{7.3.20}$$

with dispersion matrix

$$\begin{aligned} D(\mathbf{a}_A) &= (\Sigma \mathbf{Z}' \mathbf{\Omega}^{-1} \mathbf{Z})^{-1} \sum_{i=1}^{N} \sum_{j=1}^{N} \mathbf{Z}_i' \mathbf{\Omega}^{-1} E(\boldsymbol{\epsilon}_i \boldsymbol{\epsilon}_j') \mathbf{\Omega}^{-1} \mathbf{Z}_j (\Sigma \mathbf{Z}' \mathbf{\Omega}^{-1} \mathbf{Z})^{-1} \\ &= (\Sigma \mathbf{Z}' \mathbf{\Omega}^{-1} \mathbf{Z})^{-1}. \end{aligned} \tag{7.3.21}$$

Since both the LS- and Aitken estimators are unbiassed, we shall compare the dispersion matrices for these two estimators. The LS-variances are here found to be greater than or equal to the Aitken variances. This assertion is correct if it can be shown that the difference between the matrices (7.3.15) and (7.3.21) is positively semi-definite. A simple proof for this can be given as follows.

Consider the difference between the two estimators (7.3.12) and

---

[1] This result is presented earlier in Norlén (1966).

[2] The problem of estimating a set of linear regression equations is considered by Zellner (1962, 1963). Zellner estimates the dispersion matrix of the disturbances with the empirical dispersion matrix obtained from the ordinary least-squares method applied to each relation separately. In a footnote Zellner mentions the possibility of using the above algorithm (7.3.18) (idem, 1962, p. 363). For a proof of the convergence of the iterations, see Norlén (1966). In Kmenta & Gilbert (1968) the above-mentioned estimation in two steps and the iterative estimator are studied by the Monte Carlo method.

(7.3.19). This stochastic variable has expectation equal to zero and dispersion matrix, which from some simple calculations is got as

$$D(\mathbf{a}_{LS}-\mathbf{a}_A) = D(\mathbf{a}_{LS}) - D(\mathbf{a}_A). \tag{7.3.22}$$

Since a dispersion matrix is of necessity positively semi-definite, it follows that the difference in the right side of (7.3.22) is also positively semi-definite, which is what was to be shown.[1]

What has been stated above does not conflict with the known circumstance that the three estimators described coincide if the dispersion matrix of the disturbances is diagonal and/or the relations have the same regressors, see e.g. Goldberger (1964) or Theil (1971).

7.4. Survey errors

The existent need for estimates of error characteristics for testing purposes is described in Chapter 5. In particular, we were faced with the need for a study of the following two error terms: the propagated error (5.2.19) and the measurement error in (5.2.22). This section is devoted to a few summarizing accounts of the problems connected herewith.

If for the moment we neglect the propagation effect in (5.2.19), both problems may be formulated as a study of the errors $\boldsymbol{\upsilon}$ in

$$\mathbf{a} = \boldsymbol{\alpha} + \boldsymbol{\upsilon}, \tag{7.4.1}$$

where $\mathbf{a}$ is an obtained estimate or observation and $\boldsymbol{\alpha}$ is the correct value.

In the first place we need to calculate the expectation of the error

$$E(\boldsymbol{\upsilon}) = E(\mathbf{a}) - \boldsymbol{\alpha}. \tag{7.4.2}$$

In the next step we need to calculate the dispersion matrix of the error

$$D(\boldsymbol{\upsilon}) = E(\mathbf{a}\mathbf{a}') - E(\mathbf{a})\,E(\mathbf{a}'). \tag{7.4.3}$$

These simple formulations of survey characteristics with concomitant needs for estimation entail many problem-complexes that go deep and include most of the problems connected with surveys.

We distinguish between the following three sources of error:

(i) sampling variation
(ii) measuring instruments
(iii) reading off measuring instruments.

The measurement error (5.2.22) includes the sources of error (ii) and (iii), whereas the propagated error (5.2.19) includes, more or less, all three

[1] For a proof in the case of two relations, see e.g. Goldberger (1964). For large samples, the result is obtained in view of the fact that ML-estimators are asymptotically efficient; see Cramér (1945).

sources of error. We make the following comments concerning the listed points.

(i) *Sampling variation.* In the field of statistics the sampling variation of estimators has hitherto attracted great interest, and in this connection it is often assumed that the observations can be obtained without the measurement errors (ii) and (iii). If through a simple illustration we latch up to Example 7.3 in the previous section, we get the LS-estimator $\mathbf{a}_{LS}$ of $\boldsymbol{\alpha}$ from (7.3.12). It proves that for this estimator the expectation of the error $\boldsymbol{\upsilon} = \mathbf{a} - \boldsymbol{\alpha}$ is equal to zero and the dispersion matrix of the error is equal to (7.3.15). It is characteristic for statistics of this kind that the properties are for the most part obtained from deductions made in accordance with formal logic and mathematical statistics.

(ii) *Measuring instruments.* In practice, interest appears to be focussed on the problem of choice of variables, i.e. the choice of the rule or rules of measurement with reference to the real objects.[1] The choice will of course be decisive where the specific empirical measurements or indicators belonging to or most suitable for the constructions of the variables in the model are not immediately given. One characteristic of this problem is that in their main features solutions must be based upon objective-logical arguments; cf. the concept "content-universe" in the sociological Guttman-scale technique.[2] Section 9.6 deals with the problem of measuring variables and with attempts to arrive at solutions of this problem with the help of factor analysis.

(iii) *Reading off measuring instruments.* The third ramifying problem is constituted by errors in measurement incurred in connection with the collecting of primary observations. Thus for its solution this problem calls for detailed studies of the actual process of observation through control measures and evaluations.

Considerable efforts have in various contexts been made with the aim of attempting to give a total view of the problem inherent in survey design. The concept total survey design has included the content and import of the task and stressed the total approach.[3] Among other things, efforts have thus been made to integrate the three errors (i)–(iii) in a common frame of reference. The attempts at the U.S. Bureau of the Census give illustrative examples of calculating characteristics (7.4.2–3) obtained from divisions of these in sums of terms corresponding to different contributions to the total error.[4]

---

[1] Compare with the following similar statement from an econometrician: "It must be emphasized that by far the most time-consuming part of the work, and also the one that is least capable of being applied in detail elsewhere, is the estimation of the variables and not, as some may think, the estimation of the parameters in the relationships." (Stone 1954, p. 406).

[2] For a review of Guttman-scales, see Himmelstrand (1961).

[3] The argument is borrowed from Dalenius (1971), where further references are given.

[4] For discussions of these procedures, reference is made to Hansen et al. (1961, 1963).

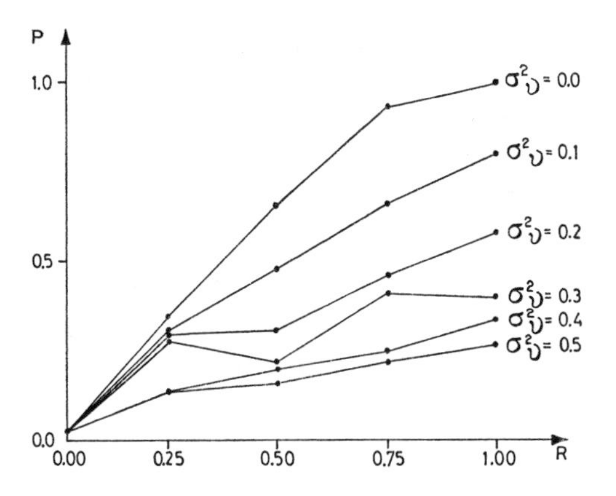

Fig. 7.5. Estimate of the probability P of getting the correct structure with the exploratory technique described in section 7.2 from Monte Carlo experiments with the system (7.2.23). The probability P is given as a function of the degree of explanation R defined by (7.2.26) for the case in which the eight observations of the six variables do not suffer from measurement errors ($\sigma_v^2 = 0$), and for five different variances of measurement errors ($\sigma_v^2 = 0.1$, 0.2, 0.3, 0.4 and 0.5).

Suitable models for the analysis of variables and estimators of the kind used by e.g. the U.S. Bureau of the Census are regarded as requisite aids in preparations for investigations of the consequences of the survey errors. If such "error models" can be successfully applied and estimates (7.4.2–3) are obtained, the remaining problems are more of a technical nature. Thus, for example, the propagated error (5.2.19) may be investigated with sensitivity analysis, described in section 6.4, giving the effect on model behaviour caused by errors in the parameter values and in the predetermined variables.

In the following exemplification in connection with the calculation of the effect of a source of error in surveys we are latching up with the technique for exploration in the real system presented in section 7.2. The sensitivity of this technique with reference to measurement errors in the observations is here investigated.

Example 7.4. Let us consider the exploratory technique presented in section 7.2 for the selection of an econometric structural form proceeding from a set of observations arranged in two categories: endogenous and predetermined variables. In Example 7.1 this technique was investigated with the Monte Carlo method for the case in which the observations did not suffer from measurement errors.

For this Monte Carlo study the same experimental design with the same relation structure (7.2.23) as in Example 7.1 is used. Instead of using correct values one observes the six variables with normally distributed measurement errors which are independent of each other and independent of the correct values. It is further assumed that the measurement errors have the mean value 0 and variances which in due succession are put equal to 0.1, 0.2, 0.3, 0.4 and 0.5. These measurement errors are assumed to be not known: the technique is applied as if the observations were without measurement errors. The results of the studies made are shown in Fig. 7.5. For comparison, the results emerging in Example 7.1, where the observations did not suffer from measurement errors, are repeated. The results show—in accordance with what is to be expected—that the frequency of successful identification of the correct structure diminishes when the magnitude of the measurement errors increases, and in five cases of six when the "degree of explanation" in the system diminishes.[1]

[1] For a more general study of the effects of measurement errors on causal inferences, see Norlén (1972). The study includes, on the one hand, the case when the measurement errors are neglected and, on the other hand, when the case in which the procedures studied are to a varying extent adjusted with respect to measurement errors.

Part II: Applications

Chapter 8

Alignment of the applications presented

8.1. Introduction

This chapter will be devoted to a few summarizing comments on the alignment given to the two applications presented in Chapter 9 and in the Appendix.

Briefly, in the theoretical part an approach has been outlined for work with computer simulation models that corresponds with generally accepted statistical norms for the treatment of parametric models. It has been shown that the two fields of estimation and testing dealt with in statistical theory and tried out in applications are well-adapted for modelling of the kind indicated here; cf. Chapters 5–7.

In the methodology devised there are inbuilt endeavours aiming at providing a common frame for model construction in several fields of application. An ideal in this respect is provided by e.g. regression analysis. The development of the general form and structure for regression-analytic models with appurtenant statistical distribution assumptions has made it possible to develop a method extending over the boundaries between different fields of application.

Seen against this background it is not necessary, for a consolidation of the present methodology, to treat in any great detail—as a first step in the application—those parts of the routine in the testing of hypotheses that refer to the proved fields of estimation and testing.

8.2. Comments on the following applications

In Chapter 9 a model of an educational system is presented, and in the Appendix an earlier constructed general model of a system of communications is analysed. Actually, however, the work in the Appendix comes, as a matter of chronological sequence, before the work in Chapter 9.

Although the two models in part cover different empirical fields, they represent in the main the same mode of attack, in which connection simulation enters as a tool for the obtaining of syntheses; cf. Chapters 1 and 2. Thanks to this, it has been possible in the work with the two models to place the main emphasis, and this with advantage, differently. In the next chapter the main stress is placed on investigating the possibility of formulating a model in conformity with the frames set up in the theoretical part. In the Appendix the work is concentrated chiefly on the problems of model analysis that arise when one is working with the present types of model.

In the modelling work of which an account is given in Chapter 9 attention is restricted to the question of the possibility of realizing the approach and to the therewith closely connected question of empirical meaningfulness in the attempt to achieve a valid model. These two questions resolve themselves into the need for an account of a functioning model which in itself has meaningful model categories permitting of, amongst other things, the fitting of theoretical results and emprical generalizations into the frame of the model. It is also relevant here, as in applied science, to read into the concept empirical meaningfulness significances or imports referring to the possibilities of using the model in the task set.

In the modelling work reported in the Appendix the theme is the understanding of a developed model. As has been mentioned, this task takes as its point of departure an already constructed general model. The task implies, briefly, through analysis of simulated data from the model getting knowledge of the latter to help in developing both its theoretical and empirical parts in the direction of adaptation to selected particular systems.

Chapter 9

Simulation modelling of education[1]

9.1. Introduction

In this chapter our interest is directed towards the application of the principles adduced for the construction of models in special educational contexts.

The educational and learning processes which occur include a complex and dynamic interaction between teacher, pupils and textbooks etc. A number of variables are of course involved in the processes, and this implies that in connection with the statistical processing of relevant problems in different reports and experimental operations there are sometimes regularly recurring features that are difficult to identify. The field is thus suitable as an object for studies pertaining to and inherent in the method presented in the foregoing.

Thus an endeavour to achieve a total breaking down of the whole or parts of the educational process in order to get at the individual causal sequences presents itself as a possibility if this enables us to get at and constructively process the results of such an analysis with a model. The whole idea is based on the material prerequisite that the result arrived at by the model shall be examined through simulation on a computer.

From other points of departure and from especially empirical results the questions of method as regards suitable approaches for the tackling of research problems in education have been taken up in educational science. If the problem situation is to be described in a simplified way the matter may be expressed thus: that in connection with empirical studies much interest has been given to frame and target variables at the expense of process variables, i.e. those variables that occur in connection with or are caused by the actual educational situations (Dahllöf 1971, Lundgren 1972).

That a study of process variables may have decisive importance for an understanding of the course of education and its effects is shown by, among other things, research on the so-called differentiation question,

[1] This chapter is based on Norlén (1970).

where one study shows that the values for the process variables give sufficiently marked differences in different groupings of pupils (Dahllöf 1967).

The present approach is one way—among several possible ones—to tackle educational problems with simulation. In other ways simulation, which is based on a division of the educational process into separate parts, has been used earlier.[1] It may also be mentioned that the method has features in common with approaches in adjacent fields.[2]

The work presented here is a limited part of a research program U & F having as its object of study education and research at the university institutes.[3] What is specific in this project is an evaluation of the simulation approach in educational science. It implies, amongst other things, an attempt to fit the method into the arsenal of other methods used in the research program; one of these latter is educational experiments in small groups. Experiments carried out in parallel with both the real system and a model of this system, simulation experiments, give reciprocal support for and further insights into the value of lines of work indicated.

An account of the aims is given in the next section.

The model is presented in a subsequent section (9.3). In that context the section contains a description of the interaction between teacher, students and textbooks etc.

In section 9.4 some preliminary simulation experiments with the model are reported. These include an investigation of the consistency of the model.

An attempt at a systematic validation is presented in section 9.5. At the present stage of development, however, the latter is possible only as a subjective validation of the model.

In its present form the model is general. This implies that the model must be adapted if it is to be possible to use it in a particular real system. In connection with this thesis the work already begun on adaptation of the model to a selected real system is described in section 9.6. From this it appears that a relatively large amount of coordinated work will be necessary. Thus a continuation should include in its assessment expected gains, which should be integrated through, inter alia, rational educational planning.

Section 9.7 reverts to the aims for the work. In this connection an illustration is given of the extent to which and the ways in which it appears possible to use the developed model for the answering of educational questions posed in planning contexts.

---

[1] An example is Carroll (1962), in which work a model to throw light upon the interaction between tuitional and pupil variables is reported.

[2] In particular, I have profited greatly from the experience gained from an analysis of the SIMCOM model constructed by Bråten (1968 *a–b*). The structure of the SIMCOM model is also partly adopted here.

[3] For an announcement of the research program U & F (*U*ndervisning och *F*orskning/ Teaching and Research/), see Wold (1970).

9.2. Aims

The motivation for using models as working instruments with a governing and controlling aim was discussed from general points of departure in Chapter 3.

The present approach has as its aim the attempt to contribute through modelling, which has been described in greater detail in the preceding theoretical part, to the answering of questions having to do with the rational planning of education.[1]

But the work has also an overlapping aim, viz, to try to devise simulation methods which may be of practical use in educational research.

The work requires an empirical field for comparative considerations of the model when measured against a modelled system. In the present attempt the model is studied at university level, to be precise, at the Statistical Institute of Uppsala University.

Where applicable, the model may be used equally well at other educational institutes or at enterprises intending to study the results of educational and training work. What we shall have to say concerning the use of the model in university contexts thus—in another design—applies also for high schools and other school forms as well.

One thing that should be required of the model is that it should yield adequate observations making it possible to assess to what extent the educational aims set up can be attained. The aim, as regards the use of the model, is thus to try to elicit facts concerning the way in which the tuition and its assimilation function in different cases and situations.

It should further be required of the model that different kinds of educational and tuitional instruments—both those at present in use and those not yet realized—shall be reproducible in it.

For the university institute, with its in part special conditions, the following instruments are listed without claim for completeness:

- Rules for enrolment (rules for access to higher studies)
- Regulations for examinations
- Textbooks etc. (content and design)
- Educational forms (combination of plenary lectures, lessons, group work)
- Educational groups (composition, differentiation and role)
- Teachers (training and tuitional skill)
- Students (previous knowledge and study situation)
- The time factor (planning of times for the tuition).

The places for and the situations of these educational instruments determine the scheduled educational situations. In an indirect way the

[1] This orientation of the work towards educational planning and the economics of education indicates a wide field of work. See Blaug (1966) and Bengtsson & Lundgren (1969) and the references indicated in these works.

instruments also determine other important learning situations, some of them generated from the scheduled tuition, in the form of self-studies, conversations and deliberations between students and teachers and among the students themselves.

So that the model shall be able to yield adequate observations of the effect of different educational instruments it ought consequently, with the aim of providing pictures of total educational effects, to give pictures of non-scheduled learning.

With this view of the task there arise a number of modelling problems of varying nature and touching upon several scientific fields. If the model is to be able to help in answering important educational questions posed, and if the modelling work is to be fruitful, it is necessary that in the programming there should be careful delimitation with clear demarcations. This can be done by indicating the frames applying for the model or, to be more precise, by indicating the view of the system that has been applied and the level of analysis that has been chosen. The description of the model in the following section (9.3) constitutes an attempt to give a relevant account of the delimitation aspect.

9.3. Construction of the model

In this section is given a description of a general model of the interaction between students, teacher and text which takes place in and follows from educational situations and other frame-conditions. In later sections we shall discuss the adaptation of the general model to a selected educational system.

The model has been given the working name SIMTEST model (SIMulation of TEacher and STudent interaction). The structure of the model follows the structure of system S described in Chapter 4. In connection with this description the construction of the model system is divided into the following three phases:

- Choice of level of resolution and classification of objects.
- Determination of input, output and state variables of the objects followed by formulation of the mechanisms of the objects.
- Specification of the model.

As regards the level of resolution, we may first remind the reader of the relative import of the concept; cf. section 4.2. In an interpretation the institution at the level *zero* may be regarded as an indivisible unit. Inputs to the institution may be exemplified by the influx of students and monetary resources. The state can be exemplified by variables such as the number of students and teachers, premises and textbooks etc., and the resources used. Finally, the output from the institution may be described by the number of students leaving, divided into examined and non-examined students etc.

The above way of viewing the matter is motivated and is suitable for descriptive purposes and for administrative and accounting requirements. A higher level would give an all too detailed picture of the institution, a picture which would also be difficult to survey.

The *zero* level, on the other hand, is not suitable for the indication and study of structural characteristics. We therefore let the institution—which at *zero* level in the abstraction is associated with only one object—at level *one* be associated with a number of distinguishable objects. These objects are assumed to be divisible into three classes, which for the sake of simplicity are referred to with natural concepts, viz, the object-classes

teacher, *student* and *text*. (9.3.1)

The system S thus consists of a number of objects from each of these three classes. The two classes *teacher* and *student* are to begin with to be understood in the intuitively comprehended sense. The objects in these two classes have of course many features in common. In considerations of such common or joint characteristics the class designation *person* is used. By the class *text* is meant the collection of distinguishable objects that compose the picture of the available textbooks etc.; an object in *text* may be e.g. a textbook for a course.

This view of the system will be used and developed. The desideratum of achieving accuracy in the description might at this stage lead us to attempt to go still deeper into the field and to introduce another level. For the class *person* this would entail e.g. an intraperson level, which would even lead to human physiological considerations. There is much to suggest that this is not suitable. In its turn it would probably entail all too great demands on the content of the theories in the social sciences involved. It would, moreover, in the construction of the model give rise to difficulties in keeping control of the whole network of interactions.

The objects in the classes (9.3.1) interact with each other and with an environment. For the time being it is sufficient to let the interactions with the environment consist only of inputs to *teacher*-objects. These inputs from the environment are prescriptions for *teacher*-objects to try to give outputs to *student*-objects. These prescribed outputs constitute the summation of different scheduled educational situations and are referred to as formal outputs. Other, not prescribed, outputs from objects are called informal outputs.

The possible interactions, the dyadic joint actions, between the objects can therewith be illustrated as in Fig. 9.1 in the simple case where the system S contains two objects from each of the three classes.

The presentation so far is dependent upon and directed towards expected possibilities of choosing variables and mechanisms for the objects.

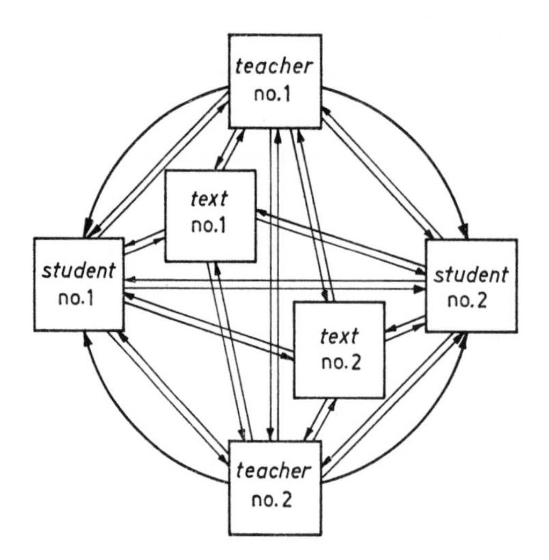

Fig. 9.1. The possible interactions in the system with two objects in each of the three object-classes. The thick arrows designate formal outputs from *teacher*-objects.

Before discussing the application of theories it will be as well to introduce some further concepts and classifications. For pedagogical and programming reasons it is practical to use the process concept. This concept is not described in the presentation of system S in section 4.2 and is not to be confused with the process concept in the SIMULA language. What we will call processes are procedures in the computer program.[1]

One may say that a modelling mechanism expresses and gives results from the moment when the object has been involved in and gone through one or several processes. At every point of time every object undergoes a process. This process is then said to be in an active phase. The other processes are here said to be in the passive phase. In the active phase a process may be connected with the active phase in a process with another object, i.e. there is an interaction. This connecting of the active phases occurs through the discharge of output variables from one of the objects which are the same as the input variables in the receiving object; cf. (4.2.8). This usage is consistent with Bråten (1968*c*).

For the objects in the class *person* we set up the following five processes:

receive, *send*, *adapt*, *rest* and *decide*. (9.3.2)

The transitions between the active phases in the *person*-processes are shown in Fig. 9.2.

The objects in the class *text* may be suitably regarded as reduced *person*-objects with the following three processes:

receive, *send* and *rest*. (9.3.3)

[1] The orientation in this work towards a particular process concept—often in educational contexts referred to as the problem of the "black box"—has relevance to the current discussion in educational research; see e.g. Dubin & Taveggia (1968).

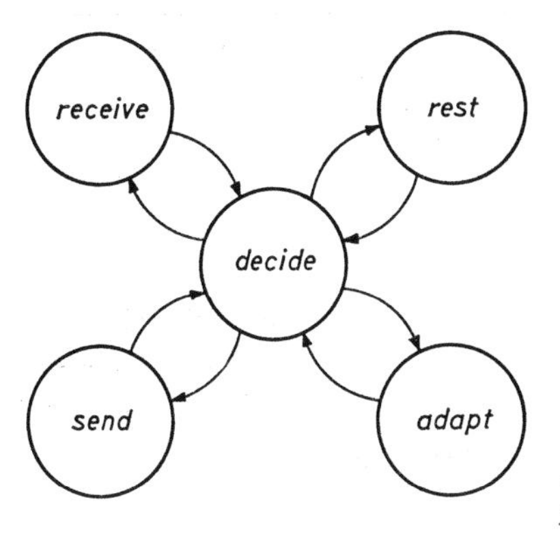

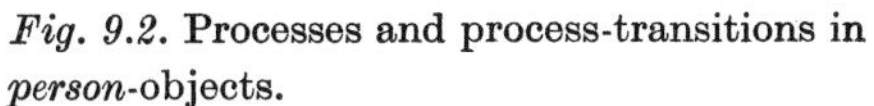

Fig. 9.2. Processes and process-transitions in *person*-objects.

A *text*-object is in the process *rest* when no *person*-object is in the process *send* and has suitable values for its state variables. In this connection the *text*-object goes through its processes in the order indicated in Fig. 9.3.

When an object goes through its processes the state of the object is thus changed, and output variables will possibly be formed and given out. Every part-process refers to a subset of the object's own state variables and, in interaction phases, to state variables in linked objects. The change that occurs is a function, (part-)mechanism, of involved variables and mechanism-parameters.

In Fig. 9.4. is given a clearly set out structural description of the objects, and in this connection the processes are assumed to begin at the point of time τ and end at the point of time $\tau + \Delta\tau$. As has already been indicated, the process *decide* is activated in *person* after the termination of one of the other processes.

From the information viewpoint we find from a consideration of the (part-)processes described in Fig. 9.4 that M_4 and M_6 in *send* and M_{14} in *decide* affect the object's choice among information alternatives. The part-process M_1 in *receive* may be described as the object's information-collecting body. The part-processes M_3, M_8, M_{10} and M_{12} imply a change in the state of the object as a consequence of activities performed in the processes *receive, send, adapt* and *rest* respectively. These part-processes may thus be regarded as the object's bodies for handling and processing information. The remaining part-processes are the picture of the temporal aspects of the activity.

After these preliminary delimitations of the system, consisting of the tenta-

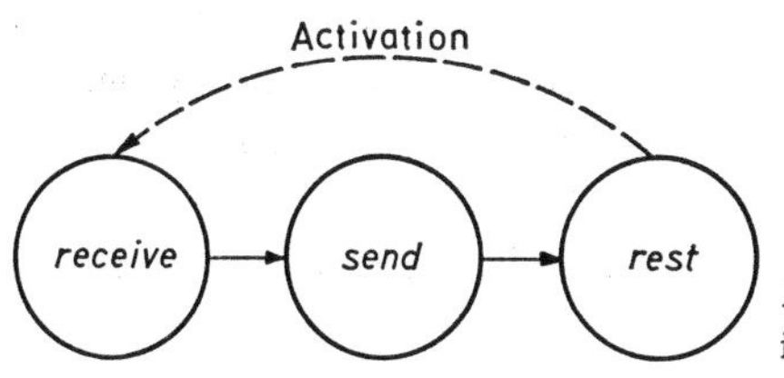

Fig. 9.3. Processes and process-transitions in *text*-objects.

Process	At point of time	Part-process	Symbol for part-process	Special characteristics for objects in the class	
				teacher	*text*
receive	τ	Receive input	M_1		
		Determine reception time $\Delta\tau$	M_2	Conditioned	
	$\tau+\Delta\tau$	Change state	M_3		Not applicable
send	τ	Seek receiver	M_4	Conditioned	Receiver is the *person* from whom the input was got
		Determine seeking time $\Delta\tau'$	M_5	Conditioned	$\Delta\tau'=0$
		If receiver not found $\Delta\tau=\Delta\tau'$ and go to M_8 otherwise			Not applicable
	$\tau+\Delta\tau'$	Form and give out output	M_6	Conditioned	
		Decide sending time $\Delta\tau''$	M_7	Conditioned	
		$\Delta\tau=\Delta\tau'+\Delta\tau''$			
	$\tau+\Delta\tau$	Change state	M_8		Not applicable
adapt	τ	Determine adaptation time $\Delta\tau$	M_9	Conditioned	This process does not exist
	$\tau+\Delta\tau$	Change state	M_{10}		
rest	τ	Determine resting time $\Delta\tau$	M_{11}	Conditioned	The object is in this process unless some *person* is sending to the object, when there is transition to *receive*
	$\tau+\Delta\tau$	Change state	M_{12}		
decide	τ	Determine deciding time $\Delta\tau$	M_{13}	Conditioned	This process does not exist
	$\tau+\Delta\tau$	Determine next process	M_{14}	Conditioned	

Fig. 9.4. Description of function of the part-processes. The conditions applying for *teacher*-objects derive from the prescriptions for the behaviour obtained from the environment to the system.

tive formation of concepts and classifications, the formulation of the model proper begins. The construction of the system S has now gone forward so far that the next step is the determination of variables and formulation of mechanisms.

In the present version of the model the theoretical basis consists chiefly of assumptions concerning "satisficing" behaviour. According to the behavioural principle of satisficing, individuals try to act in such a way that certain variables are kept within certain limits determind by aspirations, expectations etc.[1] In a free interpretation of this paradigm the main line in the approach is summed up in the following way:

The persons in a system are carriers of a variable target-structure, which is

[1] See e.g. the review article by Simon (1959).

Object	Type	Symbol	Abbreviation of	Kind
Person	Parameter	*ID*	IDentification	Integer
	State variable	*AS*	ASpiration	Boolean
	State variable	*ER*	Expected value of Reward	Boolean
	State variable	*SA*	SAtisfaction	Boolean
	State variable	*IN*	INterest	Boolean
	State variable	*KR*	Knowledge, Retained	Real number
	State variable	*KT*	Knowledge, Total received	Real number
	State variable	*GM*	Informal Group Members	Object set
Text	Parameter	*ID*	IDentification	Integer
	Parameter	*CO*	COntent	Real number

Fig. 9.5. Parameters and state variables in the objects.

modified gradually and after experiences of earlier action. The behaviour is not governed according to the principle of maximization—it aims at satisfying needs.

In the present attempt the parameters and state variables indicated in Fig. 9.5 are therefore taken up to be included in the objects. The parameter *ID* meets the requirement of keeping in mind the individual objects in the object-classes. The few one-dimensional and sometimes even only dichotomous variables of course imply a crude description, though at the present stage they meet the need for simplicity of expression. We shall in section 9.6 revert to desired extensions of the number of variables with, inter alia, several dimensions for each variable.

Inputs and outputs are indicated with the variables in Fig. 9.6. The variable *MTY* is in this connection used for description of the kind of information. From Fig. 9.1 we learn that inputs/outputs may be of four kinds: informal input/output between *person*-objects ($MTY=1$), formal output from *teacher* to *student* ($MTY=2$), informal output from *text* to *person* ($MTY=3$) and informal output from *person* to *text* ($MTY=4$).

In section 4.2.2 is given an account of the way in which the objects are represented in the computer program in the form of separate program blocks. If we revert to the system exemplified in Fig. 9.1, then the structuralization of the objects may be illustrated in the manner emerging from Fig. 9.7.

Symbol	Abbreviation of	Kind
SID	Sender's IDentification	Integer
MTY	Message TYpe	Integer
MTI	Message TIme	Real number
MCO	Message COntent	Real number

Fig. 9.6. Variables in inputs and outputs.

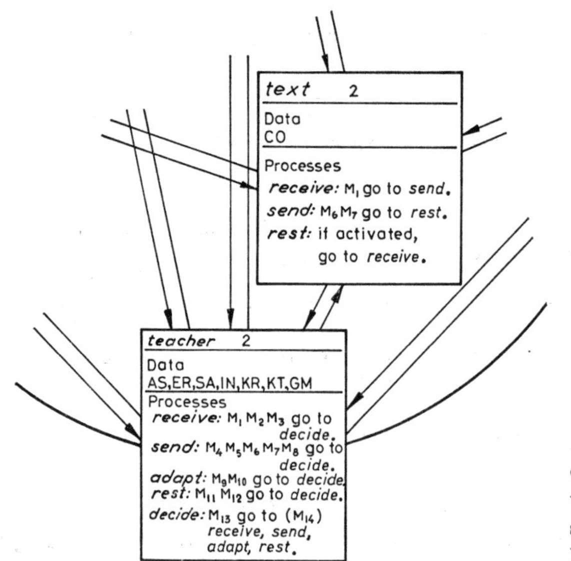

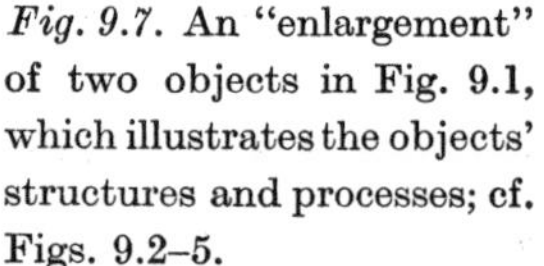

Fig. 9.7. An "enlargement" of two objects in Fig. 9.1, which illustrates the objects' structures and processes; cf. Figs. 9.2–5.

To sum up the main points in the train of thought, there is in every object at every point of time a set of values for the state variables and parameters. Further, at every point of time one of the processes is in an active phase which in mechanisms changes the value for the object's own state variables. If the object is in the process *send*, then the object also collaborates in changes of state in the object or objects existing in the process *receive* and receives the output discharged. In other words, the objects are presumed to exist parallel in time. As regards the programming language SIMULA and the possibilities for this language to describe and quasi-parallel process objects going through processes at the same time, see section 4.2.2.

As basic framework or skeleton for the formulations of mechanisms Fig. 9.8 is used. In this case the "behaviour" designated in the Figure is logically associated with the five processes (9.3.2).

As indicated in the Fig. 9.8 the variable Satisfaction (SA) depends on the variables Expected value of Reward (ER) and ASpiration (AS). The value of a *person*'s SA is recalculated each time the *person* enters the process *adapt* by evaluation of the logical expression

$$SA = \{ER = AS\}, \tag{9.3.4}$$

i.e. the *person* "compares" his actual ER with his actual AS. Two different kinds to satisfaction are thus possible: a *person* has either both ER and AS equal to "true" or they are both equal to "false".

Further, as also indicated in the Fig. 9.8, the variable SA is used in such a way that the variable is "negatively correlated" with "behaviour". One part-mechanism, for example, may thus be found in the process *decide*

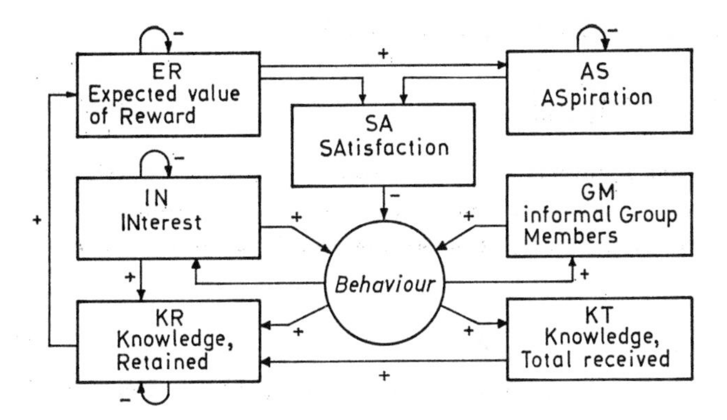

Fig. 9.8. Model of adaptively motivated behaviour.

determining the next process (*receive, send, adapt,* or *rest*), giving a smaller probability for the entrance to the process *send* if the *person*'s *SA* is equal to "true" than if his *SA* is equal to "false", *ceteris paribus.*

In the more detailed specification of the mechanisms some theoretical syntheses and empirical generalizations included in the SIMCOM-model analysed in the Appendix have been used, above all in the processes *receive* and *send*. Thus the elements fitted in from this model refer chiefly to man as an information-processing system. In the Examples 9.1–3 are illustrated some part-mechanisms in the model.

Example 9.1. In the process *receive* in *person IN* and *KT*, inter alia, are changed when an input is received with the variable-values *SID*, *MTY*, *MTI* and *MCO*; cf. Fig. 9.6. The probabilities p and q are read first from

$$\begin{pmatrix} p \\ q \end{pmatrix} = \overset{\displaystyle MTY = \atop \begin{matrix} 1 & \;\; 2 & \;\; 3 \end{matrix}}{\begin{bmatrix} \alpha_{11} & \alpha_{12} & \alpha_{13} \\ \alpha_{14} & \alpha_{15} & \alpha_{16} \end{bmatrix}}. \tag{9.3.5}$$

After this the Boolean variable *IN* goes through a step in the Markow scheme

$$\begin{array}{l} \\ \nearrow \\ \text{true} \\ \text{false} \end{array} \begin{array}{c} \begin{matrix} \text{true} & \text{false} \end{matrix} \\ \\ \begin{bmatrix} p & 1-p \\ q & 1-q \end{bmatrix} \end{array} \tag{9.3.6}$$

The constant α, measured in knowledge per time-unit, is then calculated from

$$\alpha = \begin{pmatrix} \alpha_{17} \\ \alpha_{18} \\ \alpha_{19} \\ \alpha_{20} \end{pmatrix} \text{if } MTY = \begin{pmatrix} 1 \text{ and } KR > \text{sender's } KR \\ 1 \text{ and } KR \leqslant \text{sender's } KR \\ 2 \\ 3 \end{pmatrix}. \tag{9.3.7}$$

The values for KR and KT will be increased by

$$\begin{pmatrix} \alpha \cdot MTI \\ 0 \end{pmatrix} \text{ for } IN = \begin{pmatrix} \text{true} \\ \text{false} \end{pmatrix}. \tag{9.3.8}$$

The mechanism parameters in this example are $\alpha_{11} - \alpha_{20}$. The use of the same theories that form the basis for the SIMCOM-model gives the following restrictions upon the parameter-values. On the one hand

$$0 \leqslant \alpha_{13} \leqslant \alpha_{12} \leqslant \alpha_{11} \leqslant 1$$
$$0 \leqslant \alpha_{16} \leqslant \alpha_{15} \leqslant \alpha_{14} \leqslant 1, \tag{9.3.9}$$

i.e. personal media in general attract more interest than do impersonal media, *ceteris paribus*. On the other hand, the exchange of knowledge per time-unit is greater for impersonal media, *ceteris paribus*, which gives the restrictions

$$0 \leqslant \alpha_{17} \leqslant \alpha_{18} \leqslant \alpha_{19} \leqslant \alpha_{20}. \tag{9.3.10}$$

Example 9.2. In the process *adapt* ER, inter alia, is changed and gets a new value through development of the logical expression

$$ER = \{KR \cdot \alpha_{41}/\tau \geqslant \alpha_{42}\}, \tag{9.3.11}$$

where τ is the actual point of time, α_{41} is the total duration of the simulated period and α_{42} is the knowledge requirement made at the point of time α_{41}. The expected reward ER is thus calculated by comparing the linearly extrapolated knowledge in question KR with the knowledge requirement α_{42}.

Example 9.3. In the process *rest* KR, inter alia, is changed, being reduced by

$$\min \{KR - \alpha_{61} \cdot KT, \alpha_{62} \cdot \Delta\tau\}, \tag{9.3.12}$$

where α_{61} is the fraction of KT that is always retained, $\Delta\tau$ is the duration of the process and α_{62} is a parameter, measured in knowledge per time-unit. This forgetting mechanism thus expresses that the retained knowledge KR always lies between $\alpha_{61} \cdot KT$ and KT. As long as KR is greater than $\alpha_{61} \cdot KT$, KR will diminish linearly with time.

In the present version of the model there are altogether 40 mechanism parameters in different function-expressions formed that process inputs, change states and form outputs in the processes. The above illustrations must suffice as indication of the mode of working.[1]

---

[1] Not reported are e.g. the different kinds of priority rules employed when a scheduled process is to be abrupted to enable another process to be started instead. To illustrate the need for such a priority rule mention may be made of the case when a person is in *rest* and another *person* is trying to *send* to this *person*. It then has to be "decided" whether this *person* shall enter his *receive* process.

It falls outside the frame of this work to try to give a more adequate treatment of the theoretical implications in the approach, or otherwise expressed, to try to indicate how in different ways one may connect with current research in the relevant areas. As an indication of the interdisciplinary approach the following sample from theories and research fields is given.

The survey includes overlapping theories that can form points of departure for formulations of all processes, and partial theories that can be used for the individual (part-)processes.

Among the overlapping theories is for example the field that may be referred to as theory pertaining to decisions. This includes theories concerning risk-free choice (utility maximization, rational man), theories about choice under risk (maximization of expected utility) and theory of games.[1] Also the theory for statistical decision functions can be included in this field.[2] In the criticism levelled against the theories mentioned and against their basic assumptions and normative characters theories pertaining to the concept of satisficing behaviour deriving from the theory of organization and applied here have been adduced as alternative bases for explanation.[3]

In psychology various theories of learning naturally play a central role in this connection. The factors chiefly discussed are the part-processes M_1 and M_3 in *receive* and M_6 in *send*; see Fig. 9.4 above. Further, the part-process M_{12} in *rest* may be adduced as a pure retention or forgetting mechanism.

An ideal model ought thus to contain in one or another form such cognitive aspects of the learning process as recognition, perception, conceptualization, memorizing, the drawing of inferences, solving of problems etc.

Continuing this brief survey, we may mention Homans' theory of elementary social behaviour.[4] From this point of departure it should be possible to get fresh slants on the interaction behaviours M_3 in *receive* and M_4 and M_8 in *send*. Festinger's theory of cognitive dissonance contains elements that can be utilized, especially in the process *adapt* etc.[5]

Behind the labels used for theories lie different aspects of the field.

---

[1] The two articles by Edwards (1954, 1961) give surveys of the literature up to 1960 dealing with psychological and economic theories concerning risk-free choice, choice under risk and the theory of games as well as experiments connected with these theories.

[2] The classical work is Wald (1950).

[3] In Simon (1957) satisficing and maximizing models of decision-making are compared; see also March & Simon (1966).

[4] Homans (1961). A simulation model application of Homans' theory is to be found in Gullahorn & Gullahorn (1963).

[5] Festinger (1957) and Brehm & Cohen (1962). For a simulation application of Festinger's balance and (in)congruence theories, reference is made to the model by Bråten (1968*a*) analysed in the Appendix to this monograph.

The theories are sometimes sketchy and approximative, and sometimes they cover or supplement each other. In this connection it is an urgent desideratum to make collocations from the theories presented and to try with their help to bridge over the gaps separating the numerous and sometimes differently defined and used concepts; cf. the concepts of perception and intelligence. The abstract system S and the problem defined in this connection appear to constitute a practically useful aid enabling us to set out and advance along different paths.[1]

It emerges, moreover, from the approach that the processes must be placed in their temporal context. This appears from Fig. 9.4, where the determination of times is referred to the part-processes M_2, M_5, M_7, M_9, M_{11} and M_{13}. The field lies open for parametric formulations from empirical generalizations concerning durations of time in different processes; the duration in one process may perhaps be described as normally distributed, while the duration in another process is best described by the exponential distribution etc.

In the theoretical part stress has been laid on the importance of carefully defining desired properties in the model; cf. section 5.1. Against the background of the discussion in section 2.3 work with a model specification must be based on referential interpretation rules and assumptions from used theories matched with appropriate evidential interpretation assumptions. Since in its present state the model is all too undeveloped in this respect it has been found unrealistic in this situation to bind work in the sequel to a detailed specification of the empirical part model in conformity with section 4.4.

9.4. Analysis of the model

To get a clear idea of the implications of the model it is necessary—as already explained in Chapter 6—to carry out numerical experiments, simulations, with the model. The first studies in this connection must consist of tests of the model's consistency and then a first test of its validity.

In the study, subjectively estimated values are assigned to the parameters in the model. These values are estimated on the basis of general assessments, the restrictions on the parameters obtained from the theory being duly observed; cf. (9.3.9–10). Thus a more systematic validation in connection with that described in section 5.2.3 and proceeding from a model adapted to a particular system must be postponed until the analysis of the modelled system has been taken further.

In the simulation is sketched a lesson-group consisting of one *teacher*

[1] The problem of integration of partial theories in sociology has been considered by Zetterberg (1967). He directs attention to, among other things, the already collected empirical findings concerning human behaviour in the form of 1045 numbered statements which are to be found in Berelson & Steiner (1964).

Table 9.1. *Simulation result from the SIMTEST model. The table shows the group's situation on day one and day ten*

The symbols used for the Boolean variables *AS*, *ER*, *SA* and *IN* are 1, which signifies "true" (aspires to manage the course, expects to manage the course etc.) and 0, which signifies "false" (does not aspire, expect to manage the course etc.)

		Situation at the beginning of day one							Situations at the end of day ten											
		States (=initial states)							States (=end states)							Proportion of time in processes				
																receive		*send*		*rest*
person	*ID*	*AS*	*ER*	*SA*	*IN*	*KR*	*KT*	*GM*	*AS*	*ER*	*SA*	*IN*	*KR*	*KT*	*GM*	Formal	Informal	Formal	Informal	
teacher	1	1	0	1	1	73.0	100.0	2 3	1	1	1	1	60.0	100.0	19 14 16 9 5 11 2 3	—	0.33	0.10	0.00	0.57
student	2	1	1	1	0	3.0	5.0	1 3	1	1	1	1	10.2	17.0	6 13 1 3	0.08	0.26	—	0.00	0.66
student	3	0	1	1	1	1.5	2.5	2 1	1	0	0	1	5.1	8.5	17 12 16 14 4 13 8 5	0.04	0.41	—	0.15	0.40
student	4	1	1	1	0	0.0	0.0	5 6 1 7	1	0	0	1	5.7	9.0	15 7	0.04	0.15	—	0.55	0.26
student	5	1	1	1	1	0.0	0.0	4 6 1 7	0	0	1	1	6.6	9.0	16 4	0.04	0.17	—	0.23	0.57
student	6	1	1	1	0	0.0	0.0	5 4 1 7	1	0	0	1	8.4	12.0	20	0.04	0.19	—	0.59	0.17
student	7	1	1	1	1	0.0	0.0	5 4 6	1	0	0	1	5.4	9.0	8 4 11	0.02	0.22	—	0.22	0.54
student	8	1	0	1	0	0.0	0.0	9 10	0	0	1	1	5.4	9.0	7 9 16	0.00	0.30	—	0.29	0.41
student	9	1	1	1	0	0.0	0.0	8 10	1	0	0	0	5.4	9.0	16	0.00	0.15	—	0.22	0.63
student	10	0	0	1	0	0.0	0.0	8 9	0	0	1	1	3.6	6.0	13 5	0.02	0.18	—	0.11	0.69
student	11	1	1	1	0	0.0	0.0	12	1	0	0	0	5.4	9.0	14 7 6	0.00	0.33	—	0.23	0.44
student	12	1	1	1	0	0.0	0.0	11	1	0	0	0	5.4	9.0	18 6	0.04	0.26	—	0.22	0.48
student	13	1	1	1	0	0.0	0.0	10	0	0	1	1	3.6	6.0	16 10	0.00	0.14	—	0.27	0.59
student	14	1	1	1	0	0.0	0.0		1	0	0	0	2.7	3.0		0.00	0.10	—	0.55	0.35
student	15	0	1	1	0	0.0	0.0		0	0	0	1	8.4	12.0	4 14	0.06	0.23	—	0.06	0.65
student	16	1	0	0	1	0.0	0.0	17	1	0	0	0	4.8	6.0	5	0.02	0.12	—	0.44	0.42
student	17	1	0	0	1	0.0	0.0	16	0	0	1	1	5.4	9.0	3 6 11 9	0.02	0.48	—	0.15	0.35
student	18	1	0	0	0	0.0	0.0	19	1	0	1	1	7.7	12.0	12 19	0.00	0.49	—	0.15	0.36
student	19	1	0	0	0	0.0	0.0	18	1	0	0	0	5.5	9.0	1	0.00	0.15	—	0.67	0.18
student	20	0	0	1	0	0.0	0.0		1	0	0	1	5.9	9.0	6 19	0.02	0.42	—	0.12	0.44
student	21	0	0	1	0	0.0	0.0	14	1	1	1	1	9.0	15.0	14	0.10	0.00	—	0.00	0.90
Means for *students*		1	1	1	0	0.2	0.4	—	1	0	0	1	6.0	9.4	—	0.03	0.24	—	0.26	0.47

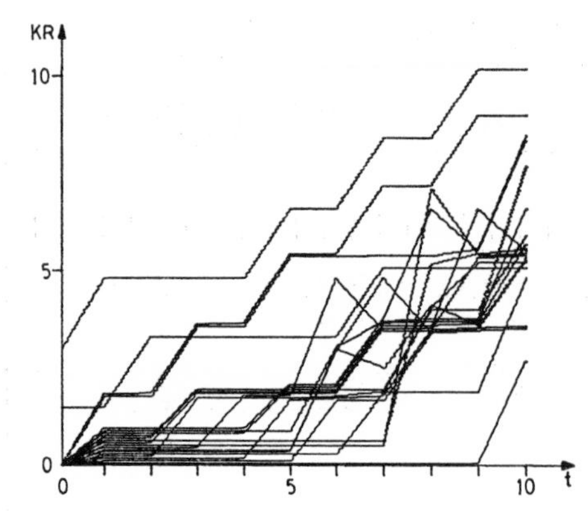

Fig. 9.9. The time-series of the variable *KR* (Knowledge, Retained) for the twenty *students*. The variables are observed at the time-points 0, 1, ..., 10 and linearly intrapolated. At time-point 1 the observations are somewhat separated from each other.

and twenty *students*. The activity of the group is simulated for ten time units (days). The *teacher* is "charged to teach (give outputs/send inputs to)" the twenty *students* at the times 0.2, 2.2, 4.2, 6.2 and 8.2 with sending time on each of 0.2 days.

For the experimental design of the study the following scheme is followed:

- Control parameters, e.g. duration of the simulation period and number of replications of the simulation.
- Model parameters, which are of two kinds, viz, global parameters, e.g. number of objects of every kind (*teacher*, *student*, *text*), and mechanism parameters.
- Object parameters (identifications), starting values for object states and starting values for the sequences of pseudo-random numbers used in the model mechanisms.
- Inputs from the environment to the model system. Among these are the instructions for *teachers* to give out outputs. Every such input from the environment is indicated with time, sender, sending time and intended receivers.
- Form for reporting of simulation results. The data that will subsequently be used for further analyses are chosen in accordance with the targets set for the study. As should have emerged from the foregoing, there are possibilities of making samples referring to system and/or subsystem aspects, state and/or flow aspects and temporal and/or spatial aspects.

Table 9.1 shows simulation results from the model with the lesson group's starting situation with the beginning of day one and situation day ten.[1] From the Table it may be read that at the end of the simulation period

[1] The first simulations were performed on the university computer CDC 3300 in Oslo and, per terminal from Oslo, on CDC 6600 in Stockholm. Simulations have since also been carried out on the university computer CDC 3600 in Uppsala.

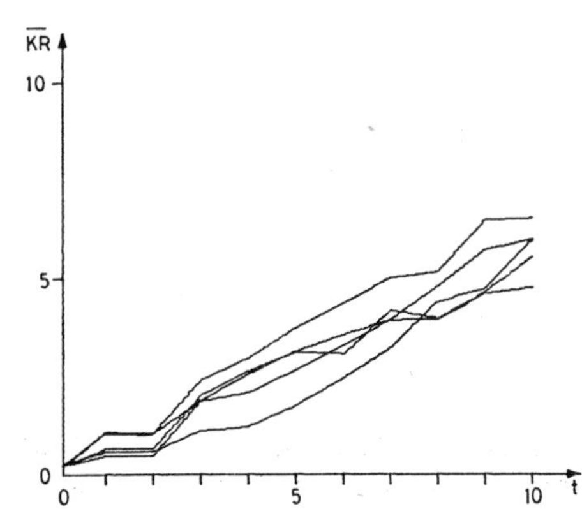

Fig. 9.10. The five time-series of means $\overline{KR}$ of the variable KR for the twenty *students* from five replications of the experiment.

the "amount learned" varies considerably among the *students*. In Fig. 9.9 these differences are further illustrated.

Five simulation runs and replications of the experiment were executed with the described design except for other sequences of pseudo-random numbers used in the model's mechanisms. In Fig. 9.10 are reported time series from the five replications. Each curve shows the development over time of the mean $\overline{KR}$ of the variable KR for the twenty *students*.

9.5. A Turing test on the model

In the study of the simulation results no direct observation of any remarkable or unexpected behavioural patterns was made. The behaviour of the *teacher* and the *students* takes a course that is reminiscent of the empirically observed behaviour at a university institute.

To ascertain whether this subjective evaluation of ours was in agreement with that of others, we constructed and distributed five questions. The

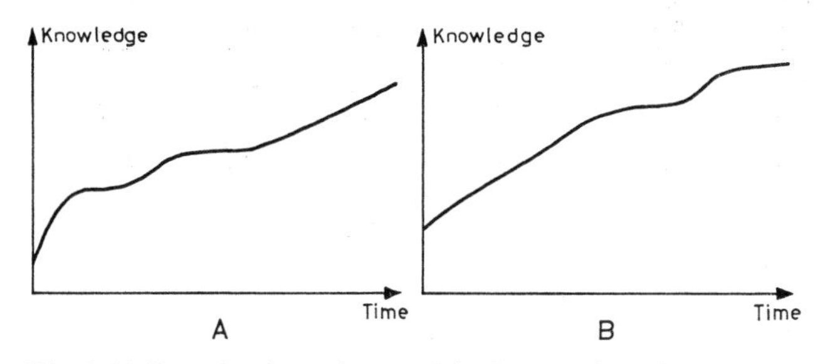

Fig. 9.11. Learning (question no. 1 in the questionnaire). Model system:
A learning curve for statistical knowledge. The curve refers to the average development of knowledge in twenty *students* during a learning period with evenly distributed tuition.
Real system:
A learning curve for typewriting. Female pupil, 32 years old. The curve refers to the number of strokes per 3-minute test at even intervals during a learning period.

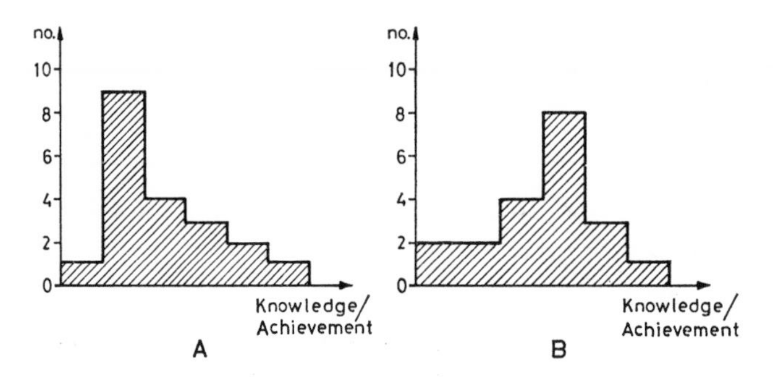

Fig. 9.12. Distribution of knowledge (question no. 2 in the questionnaire). Model system: Distribution of knowledge among twenty *students* after a learning period.
Real system:
Distribution of written result among twenty students in test on statistics.

layout of the investigation follows the method described in section 5.2.2 and based on Turing's idea. Thus in each question two kinds of information are given, viz, a collocation of observations from a simulation with the model and observations from a real system. The respondents had then to try to distinguish these two kinds of information from each other and state which was a result from the model.

The questions posed are shown in Figs. 9.11–15. The two pieces of information given in each question are marked in the Figures with A and B.

The questions were answered by 33 students at graduate level at the Departments of Statistics, Sociology, Psychology and Education in Uppsala. Table 9.2 shows the distribution of responses. The results show that the

Student	V_1	V_2	V_3
1	0.6	0.4	0.5
2	0.2	0.4	0.6
3	0.2	0.3	0.7
4	0.2	0.2	0.7
5	0.6	0.0	0.9
Mean	0.4	0.3	0.7

A

Student	V_1	V_2	V_3
1	0.5	0.2	0.7
2	0.6	0.3	0.6
3	0.6	0.2	0.6
4	0.7	0.1	0.7
5	1.0	0.0	0.7
Mean	0.7	0.2	0.6

B

Fig. 9.13. Behaviour (question no. 3 in the questionnaire). Model system:
Collocation of activities of five *students* during a learning period.
Real system:
Processing of five responses in a time-study of students. For the collection of the data the students were instructed to keep an account for one month of the respective times they allotted to different study activities.
Variables:
V_1, period of attendance for tuition/total tuition time; V_2, part of day on average for self- and groupstudies; V_3, part of day on average for other activity (amusements etc.).

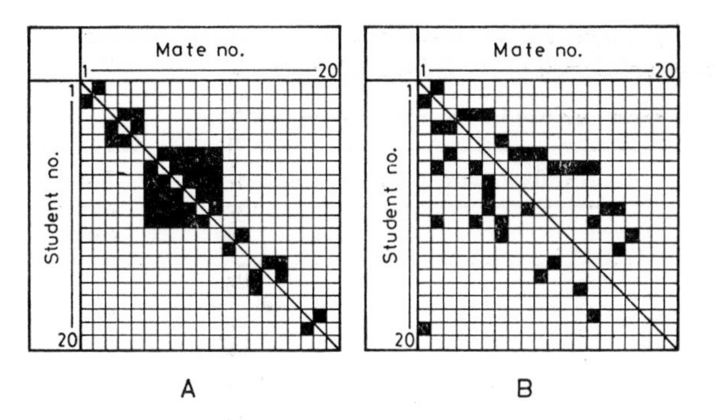

Fig. 9.14. Sociometry (question no. 4 in the questionnaire). Model system:
Twenty *students*' informal groups at the end of a learning period.
Real system:
Twenty responses from a questionnaire concerning choice of comrades for 20-point students at enrolment in statistics course autumn 1970.

total proportion of correct responses amounts to 75/165 = 0.45. As the proportion of correct response in the case of pure guessing should be on an average 0.50, this result does not motivate any change of the model.[1]

	No expectation		Expectation			No expectation		Expectation		
	No aspiration	Aspiration	No aspiration	As piration	Total	No aspiration	Aspiration	No aspiration	Aspiration	Total
No interest	6	0	6	2	14	5	2	4	0	11
Interest	0	0	4	2	6	1	0	4	4	9
Total	6	0	10	4	20	6	2	8	4	20
	A					B				

Fig. 9.15. Interest, aspiration and expectation (question no. 5 in the questionnaire). Model system:
Twenty *students*' values for the variables interest, aspiration and expectation at the end of the learning period.
Real system:
Twenty responses to a questionnaire.

[1] The remarkable fewness of correct answers to question four has a natural explanation. In this question the real system, diagram A, is characterized by very homogeneous groups. This is due to the fact that the students were requested to submit only reciprocal comrade-choices. This was in order to satisfy their desire to keep to the same group when lesson groups were formed. For obvious reasons this extra information has not been appended. The above conclusion from the investigation is not changed if this question is excluded. The critical number of correct answers for every question at the significance level 5 per cent can be approximated with

$$33 \cdot 0.5 + 1.65 \cdot \sqrt{33 \cdot 0.5 \cdot 0.5} > 21, \tag{9.5.1}$$

which exceeds the column sums in Table 9.2.

Table 9.2. *Distribution of responses to questionnaire with the questions in Figs. 9.11–15 for subjective validation of the SIMTEST model*

The correct responses to the questions are in order A, A, A, B, B

Category questioned (number questioned)	Number correct for question no.					Total number correct
	1	2	3	4	5	
Students in statistics (10)	2	4	9	0	8	23
Students in sociology (7)	4	1	3	1	4	13
Students in psychology (7)	6	6	5	1	0	18
Students in education (9)	7	6	4	0	4	21
Total (33)	19	17	21	2	16	75

9.6. Analysis of the modelled system

It is necessary to include in the model work the confrontations with a real system described in the theoretical part. In its initial stages this work has, as has been pointed out and experienced, its greatest importance for the successive revision of the model that may possibly come into the question, on the basis of comparisons with empirical evidence.

As is stated in the aim, the target for this work is to adapt the general model to the conditions prevailing at the Statistical Institution at Uppsala University.

The organization of the work of collecting and processing data within the frames indicated by the model will of course play an important role for the final result. A question of topical interest in this connection is the current introduction of automatic data processing at the university institutions. This rationalization implies in the first place the automatization of the routine work at the institutions. An interesting perspective presents itself if this rationalization is seen in connection with research on the educational process. A model of the kind here described might in this connection constitute one of the frames needed for the alignment of such research and the coordination of the results arrived at.

Since 1969 the automatic data-processing project RISK has been carried on at the Statistical Institution. Quite apart from the administrative work of registration and filing, the idea of systematization and simulation for research on conditions in the institution pertaining to educational planning etc. has been entertained from the very outset.[1]

The work so far done as regards the adaptation of the model has been done on material provided by the growing registers and files. Thus with their

[1] The target for the work is described in Information on RISK (*R*egistrering *I* *S*tatisti*K*/Registration in statistics/) (1969). See also the works by Norlén (1969) and Lindström (1970). The latest report of the structure of the register system is Jansson et al. (1971).

continuous following of different activities and developments the registers play an important role also for work in the sequel. It has, moreover, proved relatively easy to supplement this firm data foundation with investigations on special conditions.

The analytic needs as regards adaptation to real systems are described in Chapter 7. The work has hitherto embraced only the first two itemized points: description of and exploration in the real system.

The first descriptions of the system were carried out in connection with the students' enrolment for studies in statistics for the autumn term in 1969 and the spring and autumn terms in 1970. Data were collected through questionnaires concerning the students' background factors and study situation. The material thus obtained has been in part collocated and utilized for the first validation of the model; cf. section 9.5.

The following account consists of summaries of three exploratory analyses collected during the terms referred to. All these three studies may be described as analyses of variables aiming at an improvement of the empirical foundation for the state variables of the model-objects.

Exploratory analysis number 1. A large number of studies have been carried out with different kinds of correlation analysis with the aim of finding out what factors in the students have influenced success in their university studies.[1] A complex of such frequently investigated variables is constituted by so-called background factors, e.g. the students' earlier study results. A study on the effect of the background factors of students of statistics has been made which on the whole follows a traditional plan (Jansson 1971). Among the things studied e.g. with multiple regression analysis has been the correlation between the students' written results and their background.

Material from the autumn term of 1969 was processed in the following way. Data from a population of about 200 students were collected in the form of written results from about ten tests referring to part-examinations in the first term's course in statistics together with about twenty data concerning the students' background and study situation, among which the following may be mentioned.

- age, sex and civil status
- eventual experience of gainful employment
- access to educational funds in the form of study loans etc.
- new or old educational system
- eventual academic qualifications (merit rating)
- type of secondary school background
- certificates from completed secondary school education.

---

[1] A survey of studies concerning the effect of background and study situation on success in university studies is given in Ahlström (1968).

Table 9.3. *Percentual distribution of 377 responses to the question of how far student's work or career ought to satisfy specified demands*

The question was taken from Rosenberg (1957)

	Distribution of answers		
Occupational demands	Very important	Rather important	Not important
make it possible for me to look forward to a stable and assured future	44	48	8
give me the chance to earn much money	11	56	33
make it possible for me to work with people rather than things	45	32	23
give me the chance to get a leading position	6	31	63
give me the possibility of using my special gifts	38	50	12
give me the chance of working independently unencumbered by the control exercised by superiors	42	47	11
give me a high standing and a good social position	3	23	74
give me the possibility of creative activity	37	44	19
give me the chance of helping others	38	47	15
give me excitement	16	35	49

These background variables were used as regressors in linear regressions of the written results. When measured in terms of the multiple correlation coefficient the regression model was throughout found to have a low goodness of fit. The highest multiple correlation coefficient was 0.30, which was to be expected, since the individual correlations with the written results were also found to be low. The highest correlations with the written results were for marks in mathematics and average marks from secondary school education.[1]

From these results the conclusion drawn is that background factors—at least as regards some of those often used and those studied here—seem to have relatively little importance for students' success in studies of the kind here discussed.

For the modelling work the result implies in the first place that in continued empirical studies the work may with advantage be concentrated on students' orientation towards and pursuit of studies and other activities during the study period for the study and educational conditions applying.

Exploratory analysis number 2. The same questionnaire material from the autumn term of 1969 that was used in the regression analyses reviewed in the foregoing included a question concerning the extent to which an

[1] The results do not, however, show that previous knowledge is of no importance. On the contrary, some new results from diagnostic tests in the U & F research program give reason for the belief that previous knowledge in mathematics is essential for the achievement of success (Andersson 1972). It is, naturally, this actual knowledge, and not marks from school that should be reflected in the initial and predetermined values given to the knowledge variables for the *students* in the model.

occupation ought to satisfy different demands. An account of the question and the marginal distribution of the answers is given in Table 9.3. The dispersion matrix for the answers has been factor-analysed (Larsson & Olofsson 1970). The three alternative responses were quantified with 0, 1 and 2 respectively. Further, analyses were carried out with quantifications -1, 1, 2 and 0, 1, 3 respectively. In none of the cases, however, did these quantifications lead to any notable changes in the conclusion as compared with the first factor-analysis carried out.

The factor-analysis was carried out with Jöreskog's method (Jöreskog 1963), and five factors were found to be significant. The following factors were obtained:

• Security factor	assured future earn money
• People-orientated factor	work with people help others
• Interest factor	independent work special gifts creative activity
• Excitement factor	excitement
• Status factor	leading position high standing earn money

The interpretations as well as the designations of the factors follow, to a certain extent, earlier investigations of a similar kind; cf. Gesser (1967) and Gelin (1969).

It is reasonable to assume that students' orientation pattern in connection with statistical studies should resemble that described. Among other things, there emerges the plural orientation among the students. In the model are indicated the orientation and target structure of the *student*-objects with the three one-dimensional variables AS (ASpiration), ER (Expected value of Reward) and SA (SAtisfaction). The above result points to an extension of the dimension-index of these variables.

Exploratory analysis number 3. A factor-analysis of written results has also been carried out (Näsman et al. 1971). The aim of the analysis was to try to discover some underlying psychological cognitive factors that have exerted an influence in the solving of the problems.

The material included written results from three part-examinations in the autumn of 1969 covering, in all, 17 problems. Only the results from the 122 students who took part in all three examinations were included in the analysis. With a design and analysis analogous with the earlier factor-analysis three interpretable factors were obtained:

- Memory factor. In this factor are included problems that are as a rule understood as central questions by the students. Thus to solve these problems it is necessary that the courses should be carefully followed.
- Logical factor. Most of the problems belonging to this factor imply some form of probability calculation. Thus if these problems are to be solved they must be logically and systematically structured and the possible formations be formed.
- Inference-drawing factor. The consistent feature of the problems belonging to this factor is a certain form of significance testing or calculation of confidence interval. Thus if these problems are to be solved it is necessary that the students should be able to make a statistical inference from a material.

The interpretations and designations of these factors deviate somewhat from the original report. The analysis needs to be developed with, inter alia, attempts to connect "cognitive dimensions" obtained with theoretical taxonomies; see e.g. Bloom (1956) and Gagné (1965).

For the present modelling work the above results point to the necessity for an extension of the one-dimensional knowledge variables *KT* (Knowledge, Total received) and *KR* (Knowledge, Retained) in the modelling objects to comprise several dimensions.

The other analyses aiming directly at estimation of the model are to be postponed until further exploratory analyses have been carried out and the model has been revised and supplemented in connection with and in the light of the results that have emerged.

In further empirical studies the dynamic features of the model require increasing attention to the dynamic aspects of behaviour, i.e. how sets of events are distributed over time among students and teachers. Also of importance is to study how these events depend on varying environmental conditions. As an example of one such empirical study of relevance we mention the analysis of classroom interaction recently reported by Lundgren (1972).[1]

Naturally, a formalized model of the type considered here demands a specific analytic scheme for the estimation of model parameters and environmental conditions. Nevertheless, some already existing raw data are available, as in the case referred to, which may be used either in the present version of the model or in a revised version of it.

The relatively short steps that have been taken in this work to arrange and complete the model bring to the fore an important question of

[1] In Lundgren's study data obtained from tape recordings of classroom teaching were used. The communication pattern was then analysed with a category system constructed around the content of the teaching, the time factor, and the dyadic interactions. This system was complemented with a category system centered around affective components in interaction and a system categorizing the communication from a syntactical aspect.

resources. What is to be reckoned with is a coordinated work-contribution that will extend over several years and call for total planning, as is the case with major development projects with organizing and optimizing methods. An important constituent of such planning and designing of the work in the form of a data register based on automatic data processing has already been mentioned. The question of whether the research indicated here is to be linked with the developmental work for "data banks" is thus not merely a question of method, but rather a question of policy.

9.7. Use of the model

The argument about the operational use of the model, implying that different educational instruments are tried out and related to educational targets, was advanced in section 9.2 as a prime supporting motive for modelling of the kind indicated here. The relatively great resources needed to develop the model and make of it a tool has also been mentioned. We therefore conclude this chapter with some indications of the simulation experiments with the model that are foreseen as contributing in a rational planning and disposition of tuition.

In the aims set forth an account is given of some educational and tuitional instruments. If the structure for the model is considered we find that it is able to give the picture of different places and situations for these instruments. The following tabulation is based on the instruments listed in section 9.2. It is thus possible for a fictive educational case to be to some extent reproduced in the model with, inter alia, suitable values for the following:

Instrument	*Image in the model*
• Acceptance rules	Minimal requirements for starting values for the knowledge variables *KT* and *KR* in the *student*-objects.
• Examination rules	Minimum requirements for final values for the knowledge variables *KT* and *KR* in the *student*-objects, inter alia, the parameter α_{42} in Example 9.2.
• Textbooks etc.	Parameters *CO* in the *text*-objects.
• Forms of tuition	The receiver-objects in the tasks for the *teacher*-objects' transmissions. To a certain extent also start situation for the informal groups *GM* in the *persons*' objects.
• Educational groups	Same as for forms of tuition.
• Teacher	Starting values for states of the *teacher*-objects.
• Students	Starting values for states of the *student*-objects.

- The time factor — Sending times and times in the tasks for the *teacher*-objects.

Some of the institution's instruments and controllable parameters and variables can be selected in the way indicated above and occur as frame conditions in the model. The values for the parameters and variables can be changed from one simulation experiment to another. If the values are so chosen that they are suitable for the situations one wants to investigate, the effect of different situations can be evaluated with the help of corresponding simulations.[1]

In the following Examples 9.4–5 two simulation studies with the model are reported. These studies have been made with the intention of illustrating the above-adduced argument concerning the possibilities of using the model in the planning of education. The current version of the model has some of the features of a developed and applied model, but it would perhaps be pretentious to claim that the simulation results necessarily have any educational significance.

Example 9.4. In the first study we investigate an aspect of the importance of the number of lessons during an educational phase. As point of departure is chosen the situation obtaining for the simulations reported in section 9.4. The characteristic feature of this situation is that the *teacher* is "instructed to teach" the twenty *students* on five occasions at the times 0.2, 2.2, 4.2, 6.2 and 8.2 with a period of 0.2 time-units for each occasion over a learning period having the duration ten time-units (days). Five simulation experiments or replications were carried out with the model for this experimental situation with the aim in this case of obtaining knowledge of the model's stochastic properties.

The results produced in the above way are compared in this example with the results from two other, new simulation experiments. These experiments are arranged in the same way as the earlier one, except that here the *teacher* is "instructed to teach" the twenty *students* only once at the time 4.2 in the one case and ten times on the occasions 0.2, 1.2, ..., 9.2 in the other case. The duration of each lesson is retained and is thus put equal to 0.2 time-units. Also in these two cases five replications of each of the experiments are carried out.

The results as regards the *students*' "amount learned" in the two experiments are reported in Fig. 9.16. For the sake of comparison the Figure also gives the results from the preceding experiment with five

---

[1] As mentioned in the introduction, the scope of the method may be extended to other problem areas. For the modelling of e.g. internal communication in enterprises *teacher* may be taken to mean supervisor, *student* to mean worker, *text* to mean material that is worked up, and formal outputs from *teacher* to mean working instructions etc.

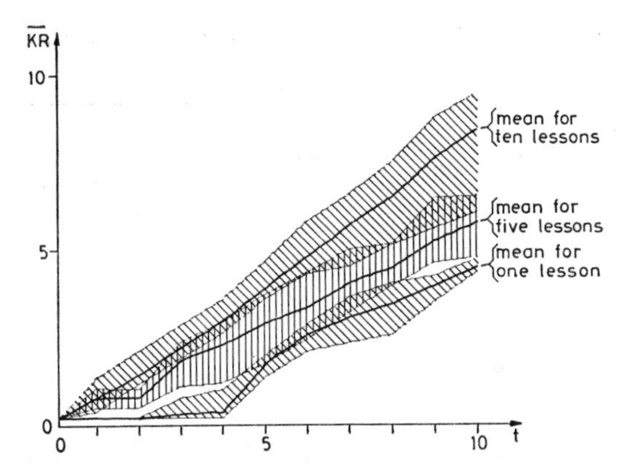

Fig. 9.16. Results from the simulation experiments with three different numbers of lessons. The Figure reports the time-series with respect to mean value and maxima and minima from the five replications for the mean value $\overline{KR}$ of the variable KR in the twenty *students* for (i) one lesson, (ii) five lessons (Fig. 9.10) and (iii) ten lessons during the simulated period of time. The areas of variation for the time-series of $\overline{KR}$ are marked with parallel lines: vertical for the experiment (ii) and inclined to the left for the other two experiments.

lessons. As might have been expected, it emerges from the Figure that the values (the effects on the knowledge-variable KR) increase with an increased number of lessons. In the Figure it is also possible to discern some occasions for the lessons; this emerges most clearly in the time-series from the simulations with only one lesson during the whole period, the curves showing considerable rises after the lesson at time 4.2.

Example 9.5. In this study reference is made to a practical/organizational question concerning the tuition at university institutions, which came to the fore in consequence of the increased influx of students in the 60's. The educational situation has implied smaller possibilities of personal contact between students and teacher. This situation, with its reduced personal contacts students/teacher is in its turn presumed to give negative effects on study-motivation and the learning process. In part for this reason the question is posed as to whether education would benefit if there were an organizational exchange of the bigger educational entities (classes, bigger groups) for small "united" study-groups under the direction of a teacher who concentrates more on the task of functioning as adviser than as direct mediator of knowledge.

With the aim of investigating the extent to which it is possible to reproduce in the model such a mode of study and then comparing the results obtained with the results from more "conventional" tuition, the two simulation experiments described in Table 9.4 have been carried out with the model.

Table 9.4. *Design for simulation experiments to compare the effects of two instructional settings*

Instructional setting	Number				Duration of each simulation
	Teachers	*Students*	Lessons	Simulations (replications)	
"Class-room tuition"					
Alt. A	1	20	1	5	10
Alt. B	1	20	5	5	10
"Studygroup"	1	4	1	5	10

The experiments with "class-room tuition" are identical with two of the three experimental series reported in the previous Example 9.4. In "class-room tuition", alternative A, the *students* have the same volume of instruction as the *students* in the "study-group". The total volume of instruction in "class-room tuition", alternative B, and the "study-group" is the same, inasmuch as the twenty *students* can be divided into five groups of four *students* each in the event of a transition to study-groups without any raising of the total educational cost.

The lesson in the experiments with the "study-group" starts at time 4.2. In other respects the experimental conditions are approximatively the same for the three experimental series. The initial situation for the experiments with class-room tuition is seen in Table 9.1. Taking this Table as point of departure the initial state for the experiment with the "study-group" was constructed as shown in Table 9.5.

Figure 9.17 shows the results produced from the model for the time-series for the mean value $\overline{KR}$ of the variable KR. As emerges from the Figure, the range of variation for $\overline{KR}$ is greater for the "study-group" than for "class-room tuition". The probable explanation for this is that the mean value $\overline{KR}$ is calculated for only four *students* in the former case, while the mean value is calculated for twenty *students* in the latter case. It further appears from the Figure that the

Table 9.5. *Initial state for the study-group*

		States at the beginning of day one						
person	*ID*	*AS*	*ER*	*SA*	*IN*	*KR*	*KT*	*GM*
teacher	1	1	1	1	1	73.0	100.0	3
student	2	0	1	1	1	0.9	1.5	5
student	3	0	0	1	0	0.0	0.0	4
student	4	0	1	1	0	0.0	0.0	3
student	5	1	0	0	0	0.0	0.0	—

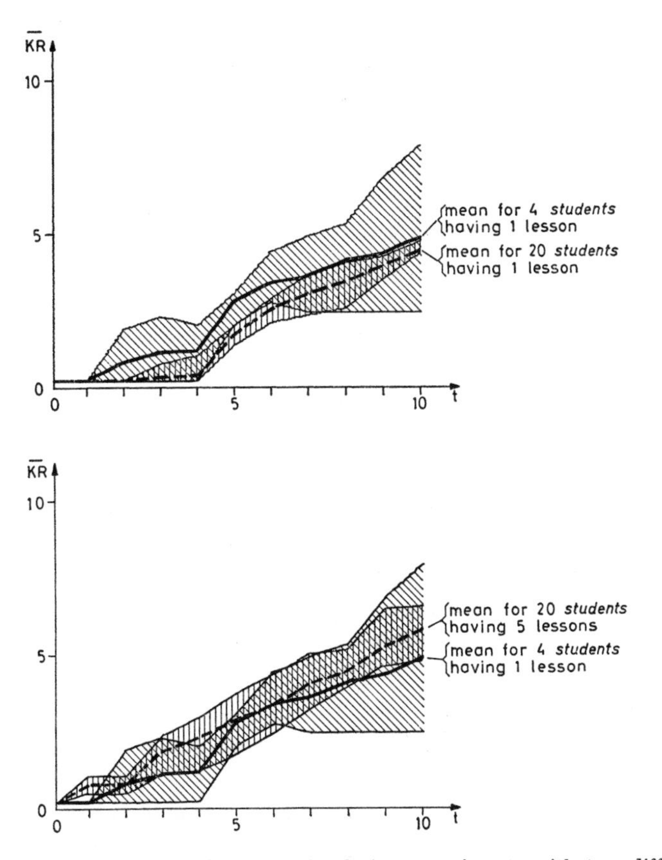

Fig. 9.17. Results from the simulation experiments with two different instructional settings. The Figure reports the time-series with respect to mean value, maxima and minima from five replications for the mean value $\overline{KR}$ of the variable KR in the *students* for (i) the twenty *students* having class-room tuition and (ii) the four *students* being united in a study-group. Above (below) is class-room tuition with one (five) lesson(s) compared with the study-group. The areas of variation for the time-series of $\overline{KR}$ are marked with parallel lines: vertical for experiment (i) and inclined to the left for experiments (ii).

mean value of $\overline{KR}$ at the end of the tuition period is greater in the case of "class-room tuition" with five lessons. However, the difference is not great; as compared with class-room tuition with only one lesson the result is somewhat better for the "study-group".

It is of interest to compare this result with results from some real experimental operations. The experiments were carried out within the U & F-program and may briefly be described as follows. The students were divided into three strata according to their results in the mathematical part of a diagnostic test; cf. footnote p. 132. Each stratum was further divided into two groups: one group was given "conventional" tuition (the control group) and the other group was subdivided into small study-groups (the experimental groups). The design of the experiment is further characterized by the fact that the volume of instruction per student was equal in the two groups. The number of successful achievements proved to be about 5 per cent higher in each stratum for those students in the study-groups, on the average 80 per cent v. 75 per cent (Andersson 1972).

Corresponding to this real experiment is the model simulation experiment "class-room tuition", alternative A, v. the "study-group" reported from above. We conclude that the results from the model and the real system point in the same direction; a circumstance which very well might be coincidental in view of, on one hand, the under-developed model, and on the other hand the influence of uncontrollable factors in the real system.

An examination of the model through, inter alia, a study of the state variables of the *student*-objects (see Fig. 9.5) leads also to the observation that the evaluations of different educational situations referred to may also be aimed at affective educational targets; cf. Krathwohl et al. (1965).

Thus the planning of tuition with the help of the model implies the question of a choice between different conceivable paths for the attaining of the targets set up. A basis for an assessment of the extent to which the aims are realized in the different alternatives is obtained at the planning stage through simulation on computers. This use for the model is thus in accordance with the first objective set up for the work.

In this connection the questions of the indispensability, for the practical planning of education with simulation models, of access to computers and a knowledge of automatic data processing come to the fore. As regards the last-mentioned requirement, the requisite knowledge of automatic data processing may with advantage be taught parallel with the tuition on the use of planning models in courses on educational technology. In the case of such training, which deals with planning from theoretical and general points of departure, it is of course not necessary for the model to be adapted and validated in relation to a particular system of education. Thus with the computer program for the SIMTEST-model written down the present modelling approach constitutes an already suitably adapted and practically useful teaching aid for tuition in educational technology.

The second target set up for the work—the tackling of educational research problems with the methods indicated in the foregoing—will be reached if and when the first target is realized, or in other words when reliable predictions for different educational situations can be obtained from the model.

Finally, though the ultimate goal, in the form of a valid model, has yet to be reached, credit should be given for the incentives to further research that modelling of the kind here referred to induces. The generation of new hypotheses in the course of the modelling is one such incentive.

References

Abelson, R. P. (1968). Simulation of social behavior. In G. Linzey & E. Aronson (eds.): *The handbook of social psychology*. 2nd ed. Reading, Mass.: Addison & Wesley.

Ågren, A. (1972). *Extensions of the fix-point method. Theory and applications.* (Ph.D. Thesis.) Dept. of Statistics, Univ. of Uppsala.

Ahlström, K. G. (ed.) (1968). *Universitetspedagogik.* Stockholm: UKÄ.

Aitken, A. C. (1934). On least squares and linear combination of observations. *Proceedings of the Royal Society of Edinburgh, 55,* 42–48.

Alexander, C. (1964). *Notes on the synthesis of form.* Cambridge, Mass.: Harward Univ. Press.

Amstutz, A. E. (1967). *Computer simulation of competitive market response.* Cambridge, Mass.: The MIT Press.

Andersson, G. (1972). Rapport från undervisningsexperiment i statistik på universitetsnivå. (Manuscript.) Dept. of Statistics, Univ. of Uppsala.

Anderson, T. W. (1958). *An introduction to multivariate statistical analysis.* New York: Wiley.

Angyal, A. (1941). *Foundations for a science of personality.* Cambridge, Mass.: Harward Univ. Press. Also in Emery (1969).

Ashby, W. R. (1952). *Design for a brain.* New York: Wiley; London: Chapman & Hall, 1966.

—— (1956). *An introduction to cybernetics.* New York: Wiley; London: Clowes, 1965.

Bartlett, M. S. (1960). *Stochastic population models.* London: Methuen; New York: Wiley.

Bengtsson, J. & Lundgren, U. P. (1969). *Utbildningsplanering och analyser av skolsystem.* Lund: Studentlitteratur.

Bentzel, R. & Wold H. (1946). On statistical demand analysis from the viewpoint of simultaneous equations. *Skandinavisk Aktuarietidskrift, 29,* 95–114.

Berelson, B. & Steiner, G. A. (1964). *Human behavior. An inventory of scientific findings.* New York: Harcourt, Brace and World.

Bertalanffy, L. von (1945). Zu einer allgemeinen Systemlehre. *Blätter für Deutsche Philosophie, 18,* Nos. 3 and 4.

—— (1950). The theory of open systems in physics and biology. *Science, 3,* 23–29. Also in Emery (1969).

Birtwhistle, G., Dahl, O. J., Myhrhaug, B. & Nygaard, K. (1971). *SIMULA begin.* Lund: Studentlitteratur (in press).

Blalock, H. M., Jr. (1961). *Causal inferences in nonexperimental research.* Chapel Hill, N. C.: Univ. of North Carolina Press.

—— (1969). *Theory construction: From verbal to mathematical formulations.* Englewood Cliffs, N. J.: Prentice-Hall.

—— (ed.) (1971). *Causal models in the social sciences.* Chicago: Aldine-Atherton.

Blaug, M. (1966). *Economics of education: A selected annotated bibliography.* London: Pergamon.

Bloom, B. S. (ed.) (1956). *Taxonomy of educational objectives. Handbook I: Cognitive domain.* New York: David McKay.

Borgatta E. F. & Bohrnstedt, G. W. (eds.) (1969). *Sociological methodology 1969.* San Francisco: Jossey-Bass.

—— (eds.) (1970). *Sociological methodology 1970.* San Francisco: Jossey-Bass.

Box, G. E. P. (1964). An introduction to response surface methodology. *Technical report no. 33.* Dept. of Statistics, Univ. of Wisconsin, Madison, Wisconsin.

Box, G. E. P. & Hunter, J. S. (1961). The 2^{k-p} fractional factorial designs, parts I and II. *Technometrics, 3,* 311–351 & 449–458.
Box, M. J. & Draper, N. R. (1971). Factorial designs, the |X′X|-criterion, and some related methods. *Technometrics, 13,* 731–742.
Brehm, J. V. & Cohen, A. R. (1962). *Explorations in cognitive dissonance.* New York: Wiley.
Brownlee, K. A. (1960). *Statical theory and methodology in science and engineering.* 2nd ed. 1967. New York: Wiley.
Bråten, S. (1968*a*). A simulation study of personal and mass communication. *IAG Quarterly, 2,* 7–28. Amsterdam: IFIP Administrative Data Processing Group. Also in Stockhaus (ed.) (1970): *Models and simulation.* Gothenburg.
—— (1968*b*). *Progress report on the SIMCOM model.* Solna: Institute of Market and Societal Communication.
—— (1968*c*). *Marknadskommunikation.* Stockholm: Beckmans.
—— (1970*a*). Systemer og modellering. In A. Hörnlund, G. Ehrlemark & S. Bråten: *Systemteori.* Stockholm: Konsultkollegiet/Beckmans.
—— (1970*b*). Communication mechanisms: A synthesizing contribution to the social psychology of interpersonal and mass communication. *IMAS no. 26E1170,* Solna.
—— (1971). The human dyad. *IMAS no. 24E0271,* Solna.
Bråten, S. & Norlén, U. (1969). Simulation model analysis. *IMAS no. 11E0269,* Solna.
Bunge, M. (1959). *Causality.* Cambridge, Mass.: Harvard Univ. Press.
—— (1967). *Scientific research, I: The search for system; II: The search for truth.* Berlin: Springer.
Carrol, J. B. (1962). The prediction of success in intensive foreign language training. In R. Glaser (ed.) *Training Research and Education.* Pittsburgh: Univ. Pittsburgh Press; New York: Wiley, 1965.
Churchman, C. W. (1963). An analysis of the concept of simulation. In *Hoggatt & Balderston* (1963).
Cochran, W. G. & Cox, G. M. (1957). *Experimental designs.* New York: Wiley.
Coleman, J. S. (1964). *Introduction to mathematical sociology.* London: Collier-Macmillan.
Connor, W. S. & Zelen, M. (1959). Fractional factorial designs for factors at three levels. *Applied Mathematics Series 54.* National Bureau of Standards. Washington 25 D.C.: US Government Printing Office.
Connor, W. S. & Young, S. (1961). Fractional factorial designs for experiments with factors at two and three levels. *Applied Mathematics Series 58.* National Bureau of Standards. Washington 25 D.C.: US Government Printing Office.
Costner, H. L. (ed.) (1971). *Sociological methodology 1971.* San Francisco: Jossey-Bass.
Cox, D. R. (1958). *Planning of experiments.* New York: Wiley.
Cramér, H. (1945). *Mathematical methods of statistics.* Uppsala: Almqvist & Wiksell; Princeton: Princeton Univ. Press, 1946.
Cyert, R. M. (1966). A description and evaluation of some firm simulations. In *Proceedings of the IBM Scientific Computing Symposium on Simulation Models and Gaming.* White Plains, N. Y.: IBM.
Dahl, O. J. & Nygaard, K. (1965). *SIMULA—a language for programming and description of discrete event systems. Introduction and user's manual.* Oslo: Norwegian Computing Center.
—— (1966). SIMULA—an ALGOL-based simulation language. *Communications of the ACM, 9,* 671–678.

Dahllöf, U. (1967). *Skoldifferentiering och undervisningsförlopp.* Stockholm: Almqvist & Wiksell.

—— (1971). *Ability grouping, content validity and curriculum process analysis.* New York: Teacher College Press, Columbia Univ.

Dalenius, T. (1971). A list of ideas for research and development in the realm of total survey design and practice. *The research project "Errors in Surveys", no. 32,* Dept. of Statistics, Univ. of Stockholm.

Davies, O. L. (ed.) (1967). *Design and analysis of industrial experiments.* London: Oliver & Boyd.

Di Stefano, J. J., Stubberud, A. R. & Williams, I. J. (1967). *Feedback and control systems.* New York: McGraw-Hill.

Draper, N. R. & Stoneman, D. M. (1966). Response surface designs for factors at two and three levels. *Technical report no. 70.* Dept. of Statistics, Univ. of Wisconsin, Madison.

Dubin, R. & Taveggia, T. C. (1968). *The teaching–learning paradox.* Eugene, Ore.: Center for the Advanced Study of Educational Administration, Univ. of Oregon.

Duncan, O. D. (1966). Path analysis: Sociological examples. *American Journal of Sociology, 72:1,* 1–16.

Edwards, W. (1954). The theory of decision making. *Psychological Bulletin, 51,* 380–417. Also in Edwards & Tversky (1967).

—— (1961). Behavioral decision theory. *Annual Review of Psychology,* 12, 473–498. Also in Edwards & Tversky (1967).

Edwards, W. & Tversky, A. (eds.) (1967). *Decision making. Selected readings.* Harmondsworth: Penguin.

Efroymson, M. A. (1962). Multiple regression analysis. In A. Ralston & H.S. Wilf (eds.): *Mathematical methods for digital computers.* New York: Wiley.

Eklund, G. (1959). *Studies of selection bias in applied statistics.* Uppsala: Almqvist & Wiksell.

Emery, F. E. (ed.) (1969). *Systems thinking.* Harmondsworth: Penguin.

Evans II, G. W., Wallace, G. F. & Sutherland, G. L. (1967). *Simulation using digital computers.* Englewood Cliffs, N. J.: Prentice-Hall.

Feigenbaum, E. A. & Feldman, J. (eds.) (1963). *Computers and thought.* New York: McGraw-Hill.

Festinger, L. (1957). *A theory of cognitive dissonance.* Stanford: Stanford Univ. Press.

Fisher, F. M. (1966). *The identification problem in econometrics.* New York: McGraw-Hill.

—— (1967). A correspondence principle for simultaneous equation models. *Working paper no. 9.* Dept. of Economics, MIT.

Fisher, R. A. (1925). *Statistical methods for research workers.* 14th ed. 1970. Edinburgh: Oliver & Boyd.

—— (1935). *The design of experiments.* 8th ed. 1966. Edinburgh: Oliver & Boyd.

Fishman, G. S. & Kiviat, P. J. (1967*a*). *Digital computer simulation: Statistical considerations.* The RAND Corporation, RM-5387-PR.

—— (1967*b*). The analysis of simulation-generated time series. *Management Science, 13,* 525–557.

Forrester, J. W. (1961). *Industrial dynamics.* New York: Wiley.

Gagné, R. M. (1965). The analysis of instructional objectives for the design of instruction. In R. Glaser (ed.): *Teaching machines and programmed learning, II,* Washington, D. C.: National Educational Ass.

Gelin, G. (1969). Arbete, barn, familj och könsdiskriminering. Förvärvsarbetsundersökningen i Uppsala 1967–68. Faktoranalys 2: "Yrkesvärde"-dimensioner. Dept. of Sociology, Univ. of Uppsala.

—— (1972). Könssegrering på en lokal arbetsmarknad: Uppsala 1967–1971. (FL Thesis.) Dept. of Sociology, Univ. of Uppsala (in press).

Gesser, B. (1967). Högre utbildning och val av yrke, Del II. Dept. of Sociology, Univ. of Lund.

Goldberger, A. S. (1964). *Econometric theory*. New York: Wiley.

Gue, R. L. & Thomas, M. E. (1968). *Mathematical methods in operations research*. New York: Macmillan.

Guetskow, H. (ed.) (1962). *Simulation in social science: Readings*. Englewood Cliffs, N. J.: Prentice-Hall.

Gullahorn, J. T. & Gullahorn, J. E. (1963). A computer model of elementary social behavior. In *Feigenbaum* & *Feldman* (1963).

Haavelmo, T. (1943). The statistical implications of a system of simultaneous equations. *Econometrica, 11*, 1–12.

Hammersley, J. M. & Handscomb, D. C. (1964). *Monte Carlo methods*. New York: Wiley.

Hansen, M. H., Hurwitz, W. N. & Bershad, M. A. (1961). Measurement errors in censuses and surveys. *Bulletin of the International Statistical Institute, 38*, 359–374.

Hansen, M. H., Hurwitz, W. N. & Pritzker, L. (1963). The estimation and interpretation of gross differences and the simple response variance. *Contributions to Statistics. Pres. to Prof. M. C. Mahalanobis on the occ. of his 70th birthday*. Oxford: Pergamon Press.

Himmelstrand, U. (1961). Attitydmätning och psykologiska skalor. In Karlsson (ed.): *Sociologiska metoder*. Stockholm: Norstedts.

Hoggatt, A. C. & Balderston, F. E. (eds.) (1963). *Symposium on simulation models: Methodology and applications to the behavioral sciences*. Cincinatti: South-Western.

Homans, G. C. (1961). *Social behaviour: Its elementary forms*. New York: Harcourt, Brace & World.

IBM (1966). *General purpose simulation system/360: Application description*. H 20-0186-1.

Information om RISK (1969). *RISK-690813*. Dept. of Statistics, Univ. of Uppsala.

Jansson, L. (1971). Om tentamenssystemet vid Statistiska institutionen. Dept. of Statistics, Univ. of Uppsala.

Jansson, L., Linder, A. & Sundin, H. (1971). RISK—ett ABD-system för registrering och administration vid en universitetsinstitution. *RISK-710308*. Dept. of Statistics, Univ. of Uppsala.

Johnston, J. (1963). *Econometric methods*. 2nd ed. 1972. New York: McGraw-Hill.

Jöreskog, K. G. (1963). *Statistical estimation in factor analysis*. Stockholm: Almqvist & Wiksell.

—— (1972). A general method for estimating a linear structural equation system In A. S. Goldberger & O.D. Duncan (eds.): *Structural equation models in the social sciences. Proceedings of a conference*. Seminar press (in press).

Karlsson, G. (1958). *Social mechanisms*. Stockholm: Almqvist & Wiksell.

Kendall, M. G. (1961). Natural law in the social sciences. *The Journal of the Royal Statistical Society, Series A, 124, Part I*, 1–18.

—— (1968). Model building and its problems. In *Mathematical Model Building in economics and industry*. London: Griffin.

Kendall, M. G. & Stuart, A. (1961). *The advanced theory of statistics. Vol. 2: Inference and relationship.* 2nd ed. 1967. London: Griffin.
Kerlinger, F. N. (1964). *Foundations of behavioral research.* New York: Holt, Rinehart & Winston.
Kiviat, P. J. (1971). Simulation languages. In *Naylor* (1971).
Kiviat, P. J., Villanueva, R. & Markowitz, H. M. (1968). *The SIMSCRIPT II programming language.* Englewood Cliffs, N. J.: Prentice-Hall.
Klir, G. J. (1969). *An approach to general systems theory.* New York: Van Nostrand Reinhold.
Klir, J. & Valach, M. (1967). *Cybernetic modelling.* London: Iliffe.
Kmenta, J. & Gilbert, R. F. (1968). Small sample properties of alternative estimators of seemingly unrelated regressions. *Journal of the American Statistical Association, 63,* 1180–1200.
Kokotović, P. V. & Rutman, R. S. (1965). Sensitivity of automatic control systems (in Russian). *Automatika i Telemekhanika, 26,* No. 4, 730–750.
Kolmogorov, A. N. & Fomin, S. V. (1957). *Elements of the theory of functions and functional analysis, I.* Albany, N. Y.: Graylock.
Krathwohl, D. R., Bloom, B. S. & Masia, B. B. (1965). *Taxonomy of educational objectives. Handbook II: Affective domain.* New York: David McKay.
Langefors, B. (1966). *Theoretical analysis of information systems.* Lund: Studentlitteratur.
Laplace, P. S. (1812). *Théorie analytique des probabilités.* Paris.
Larsson, U. & Lundin, R. (1970*a*). Some methodological views on digital simulation. In H. Stockhaus (ed.): *Models and simulation.* Gothenburg.
—— (1970*b*). Relevance, validity, reliability and the use of simulation models. In W. Goldberg (ed.) (1970): *Behavioral approaches to modern management.* Gothenburg.
Larsson U. B. & Olofsson, B. (1970). Faktoranalys av Rosenbergs yrkesvärden på statistikstuderande. *RISK-700120.* Dept. of Statistics, Univ. of Uppsala.
Lindström, C. G. (1970). RISK—ett ADB-baserat system för administration, kontroll och utveckling av den edukativa processen vid en universitetsinstitution. *RISK-700204.* Dept. of Statistics, Univ. of Uppsala.
Lundgren, U. P. (1972). *Frame factors and the teaching process.* Stockholm: Almqvist & Wiksell.
Lyttkens, E. (1964). A large sample χ^2-difference test for regression coefficients. In *Wold* (1964).
Malinvaud, E. (1964). *Methodes statistiques de l'économétrie.* Paris: Dunod; Engl. transl. 1966, Chicago: Rand McNally.
Mandel, J. (1964). *The statistical analysis of experimental data.* New York: Wiley.
March, J. G. & Simon, H. A. (1966). *Organizations.* New York: Wiley.
Marshall, A. (1890). *Principles of economics.* 8th ed. 1946. London: McMillan.
Meissner, W. (1970). Zur Metodologie der Simulation. *Zeitschrift für die gesamte Staatswissenschaft, 126,* 385–397.
—— (1971). *Ökonometrische Modelle. Rekursivität versus Interdependenz aus der Sicht der Kybernetik.* Berlin: Duncker & Humbolt.
Morrison, D. F. (1967). *Multivariate statistical methods.* New York: McGraw-Hill.
Mosbaek, E. J. & Wold, H. (1970). *Interdependent systems. Structure and estimation.* Amsterdam: North-Holland.
Näsman, S. O., Thulin, H. & Widenfalk, B. (1971). En explorativ analys av

tentamenstal i statistik på 20-poängsnivån givna under höstterminen 1969. *RISK-710428.* Dept. of Statistics, Univ. of Uppsala.

Naylor, T. H. (1971). *Computer simulation experiments with models of economic systems.* New York: Wiley.

Naylor, T. H. & Finger, J. M. (1967). Verification of computer simulation models. *Management Science, 14,* 92–101.

Newell, A. & Simon, H. A. (1963). Computers in psychology. In R. D. Luce, R. R. Bush & E. Galanter (eds.): *Handbook of mathematical psychology, I,* New York: Wiley.

Neyman, J. & Pearson, E. S. (1933). On the problem of the most efficient tests of statistical hypotheses. *Philosophical Transactions of the Royal Society of London, Series A, 231,* 289–337.

Norlén, U. (1966). Studies in nonlinear estimation of multirelation models. (FL Thesis.) Dept. of Statistics, Univ. of Uppsala.

—— (1969). Synpunkter på RISK-projektet. *RISK-691204.* Dept. of Statistics Univ. of Uppsala

—— (1970). SIMTEST-modellen. Dept. of Statistics, Univ. of Uppsala.

—— (1971). The rotation problem in econometrics. Dept. of Statistics, Univ. of Uppsala.

—— (1972). Kausal inferens i mätfelssituationer. *The research project "Errors in Surveys", no. 47.* Dept. of Statistics, Univ. of Stockholm.

Orcutt, G. H., Greenberger, M., Korbel, J. & Rivlin, A.M. (1961). *Microanalysis of socioeconomic systems: A simulation study.* New York: Harper & Row.

Palme, J. (1970). *SIMULA 67. An advanced programming and simulation language.* Oslo: Norwegian Computing Center.

Popper, K. R. (1961). *The poverty of historicism.* London: Routledge & Kegan.

Robinson, E. A. (1967). *Multichannel times series analysis with digital computer programs.* San Francisco: Holden-Day.

Rosenberg, M. (1957). *Occupations and values.* Illinois: The Free Press Glencoe.

Rummel, R. J. (1970). *Applied factor analysis.* Evanston: Northwestern Univ. Press.

Russell, B. (1914). *Our knowledge of the external world.* London: Norton.

Scheffé, H. (1959). *The analysis of variance.* New York: Wiley.

Šiljak, D. D. (1969). *Nonlinear systems. The parameter analysis and design.* New York: Wiley.

Simon, H. A. (1953). Causal ordering and identifiability. In *Koopmans–Hood* (1953). Also in Simon (1957).

—— (1957). *Models of man, social and rational: mathematical essays on human behavior in a social setting.* New York: Wiley.

—— (1959). Theories of decision making in economics and behavioral science. *American Economic Review, 49,* 253–283.

Statistical Engineering Laboratory of National Bureau of Standards (1957). Fractional factorial experiment designs for factors at two levels. *Applied Mathematic Series 48.* National Bureau of Standards. Washington 25 D.C.: US Government Printing Office.

Stone, H. (1954). *The measurement of consumer's expenditure and behaviour in the United Kingdom 1920–1938.* Cambridge: Univ. Press.

Strotz, H. & Wold, H. (1960). Recursive vs. nonrecursive systems: An attempt at synthesis. *Econometrica, 28,* 417–427.

Student (W. S. Gosset) (1908). The probable error of a mean. *Biometrika, 6,* 1–25.

Theil, H. (1958). *Economic policy and forecasts.* Amsterdam: North-Holland.

—— (1971). *Principles of econometrics.* Amsterdam: North-Holland.

Tinbergen, J. (1937). *An econometric approach to business-cycle problems*. Paris: Hermann.
—— (1939). *Statistical testing of business-cycle theories, II. Business cycles in the United States of America 1919–32*. Geneva: League of Nations.
—— (1940). Economic business cycle research. *Review of Economic Studies, 7,* 73–90.
Tocher, K. D. (1963). *The art of simulation*. Princeton N. J.: Van Nostrand.
Tou, J. T. (1964). *Modern control theory*. New York: McGraw-Hill.
Tukey, J. W. (1954). Causation, regression and path analysis. In O. Kempthorne et al.: *Statistics and Mathematics in Biology*. Ames, Iowa: Iowa State College press.
Turing, A. M. (1950). Computing machinery and intelligence. *Mind, 59,* 433–460. Also in Feigenbaum & Feldman (1963).
Törnqvist, G. (1967). *TV-ägandets utveckling i Sverige*. Stockholm: Almqvist & Wiksell.
Wald, A. (1950). *Statistical decision functions*. New York: Wiley.
Wold, H. in association with L. Juréen (1952). *Demand analysis: A study in econometrics*. Stockholm: Almqvist & Wiksell; New York: Wiley, 1953.
Wold, H. (1954). Causality and econometrics. *Econometrica*, 22, 162–177.
—— (1956). Causal inference from observational data. A review of ends and means. *Journal of the Royal Statistical Society Ser. A, 119,* 28–61.
—— (1959*a*). A case study of interdependent versus causal chain systems. *Review of the International Statistical Institute, 26,* 5–25.
—— (1959*b*). Ends and means in econometric model building. Basic considerations reviewed. In U. Grenander (ed.): *Probability and Statistics, The Harald Cramér Volume*. Stockholm: Almqvist & Wiksell; New York: Wiley 1960.
—— (1960). A generalization of causal chain models. *Econometrica*, 28, 443–463.
—— (1961). Unbiased predictors. *Proceedings of the Fourth Berkeley Symposium of Mathematical Statistics and Probability Theory, I*. Berkeley, Cal.: Univ. Press.
—— (1963*a*). On the consistency of least squares regression. *Sankhyā, A 25, Part 2,* 211–215.
—— (1963*b*). Forecasting by the chain principle. In M. Rosenblatt (ed.): *Time Series Analysis*. New York: Wiley. Also in Wold (1964).
—— (ed.) (1964). *Econometric model building. Essays on the causal chain approach*. Amsterdam: North-Holland.
—— (1965). A fix-point theorem with econometric background, I-II. *Arkiv för Matematik, 6,* nos. 12–13, 209–240.
—— (1966). On the definition and meaning of causal concepts. *Transaction des Entretiens de Monaco,* 1964. Monaco: Centre International d'Étude des Problèmes Humains.
—— (1967*a*). Time as the realm of forecasting. In E. M. Weyer & R. Fisher (eds.): *Interdisciplinary perspectives of time*. New York: The New York Academy of Science.
—— (1967*b*). Forecasting and scientific method. In H. Wold, G. H. Orcutt, E. A. Robinson, D. B. Suits & P. de Wolff: *Forecasting on a scientific basis*. Lisbon: Center of Economics and Finance, Gulbenkian Institute of Science.
—— (1968*a*). Model building and scientific method. A graphic introduction. In *Mathematical model building in economics and industry*. London: Griffin.
—— (1968*b*). Ends and means of scientific method, with special regard to the Social Sciences. *Acta Universitatis Upsaliensis, 17,* 96–140.

—— (1969*a*). Mergers of economics and philosophy or science. *Synthese, 20,* 427–482.
—— (1969*b*). Nonexperimental statistical analysis from the general point of view of scientific method. *Bulletin of the International Statistical Institute, 42,* 391–427.
—— (1969*c*). Econometrics as pioneering in nonexperimental model building. *Econometrica, 37,* 369–381.
—— (1970). Statusrapport per 1.8.1970 för forskningsprogrammet Mål och Medel för Undervisning och Forskning (U & F) på Institutionsnivå, särskilt Statistisk U & F. *Meddelande från U & F nr. 1970:1,* Dept. of Statistics, Univ. of Gothenburg.
Working, H. (1933). Price relations between July and September wheat futures at Chicago since 1885. *Wheat Studies of the Food Research Institute, 9,* 187–238.
Wright, S. (1934). The method of path coefficients. *Annals of Mathematical Statistics, 5,* 161–215.
Zellner, A. (1962). An efficient method of estimating seemingly unrelated regressions and tests for aggregation bias. *Journal of the American Statistical Association, 57,* 348–368.
—— (1963). Estimators for seemingly unrelated regression equations: Some exact finite sample results. *Journal of the American Statistical Association, 58,* 977–992.
Zetterberg, H. (1967). *Om teori och belägg i sociologien.* Uppsala: Argos. Orig. ed.: On theory and verification in sociology. Tototwa: Bedminster, 1965.

Appendix

Description and models of a simulation of communication model[1]

By Stein Bråten and Urban Norlén

1. Introduction

The first parts of this paper are devoted to a brief description of a model for computer experimentation with communication systems, and of the conceptual framework for model-building applied to its development. Thereafter experiments with reduced models of this model are reported, aimed at providing descriptive tools of less complexity than that of the original.

The usage of *Gedankenexperiment* with social system models is old and widespread, while for obvious hardware and software reasons computer experimentation with complex interaction models is a recent development. Still, computer simulation in general is starting to become a widely used methodology. Starbuck and Dutton (1972) estimate that the 1921 relevant documents they found published before 1969 represent three-fourths of what has been written in the English language. The recency of socio-cultural simulations is demonstrated by a list of 25 models, selected by Gullahorn and Gullahorn (1972): With the exception of the pioneering work on social diffusion by Hägerstrand (1953), all of these models were published during the sixties. In this list they included the community referendum controversy model developed by Abelson and Bernstein as early as 1963, and the SIMCOM model—a simulation of communication model[2]

[1] The authors are grateful for computing facilities offered: The UNIVAC 1107 of the Norwegian Computing Center, Oslo, has been used for the simulations with the original model, while the reduction experiments have been carried out on the CDC 3600 of the Univ. of Uppsala, where the second-listed author has been preparing for his doctorate. The research reported on in this paper is part of a simulation model analysis and methodology program, conducted by the first-listed author, while research director of IMAS—Institute of Market and Societal Communication, Solna. He is presently associate professor (utdan. stip.) in social psychology at the Univ. of Bergen.

[2] This is a computer-operational version of a synthesizing model built by Bråten (1968*a–c*) and formulated in SIMULA, a programming language designed by Dahl and Nygaard (1965). The model-builder has had the advantage of collaborating with the latter in adapting the model to a program structure, highly suitable for human interaction system simulations due to the language characteristics.

dated 1968, which shares much of its theoretical domain with the former, and which is the subject of the descriptions and reduction modelling to follow below.

2. The model structure and the modelling framework

The above simulation of communication model has been used in a series of computer experiments in the intersection of human communication theory and theories of cognitive consistency. In this section its structure and features will be briefly described, thereby illustrating the interaction system modelling framework applied to it.

The modelling framework

This frame has been developed by Bråten (1968*a*, *c*, 1971) with the aim of facilitating

(i) integration of "islands" of theoretical knowledge relevant to the kind of human interaction system being modelled,
(ii) in the form of statements about "meaning-tight" systems of symbol processing and exchange,
(iii) allowing for computer-operational transformations, and thus for
(iv) theory-explorative simulations.

Among the basic concept designations of the frame are 'actor', 'actor resolution level', 'action program', 'process', 'field' and 'mechanism'. Their usage is partly illustrated as we go along in the description of the model structure.

An essential feature of the frame is the *dyadic* systems orientation. Any actor, be it a person, a group or a technical medium (*sic*), is defined with reference to coactors, with which he can make up interactor systems. Any symbolic action program that such an actor is capable of performing, such as production of messages, requires concurrent execution of a coactor's program, such as reception, in order to be completed. The processes that such action programs are made up of, such as selecting or paying attention to a message or selecting a coactor, contain mechanisms that operate upon actor's and coactor's state and message variables.

Actors

Actors are systems capable of executing programs that produce state changes in their internal or external environment as part of interaction sequences, involving at least one coactor.

The main actors in the model are *persons* of promotional target populations, the size of which has been respectively 400 and 100 persons in the simulations. In the following only the latter version will be referred to.

The coactors of a specific person

$$S(i, j) \quad i, j = 1, 2, ..., 10 \tag{2.1}$$

identified according to his location in the population, are other persons

$$S(i, j+1),\ S(i, j-1),\ S(i+1, j),\ S(i-1, j) \tag{2.2}$$

in his "membership group", as well as *agents* and *mass media*, promoting two incompatible choice issues, A and B, on behalf of two competing *organizations*.

Actions and messages

Actions involve state changes in the interactor system through the execution of some program, selected from the set that defines the action alternatives of the actor class.

The model actor class person may perform the actions of resting (*sic*), exchanging and receiving messages, and choosing between A and B, while the actions of the other actor classes are restricted to sending and mediating messages. Messages being composed, exchanged and processed in the simulated systems, are restricted to be supportive of either A or B, and carry evaluative and informational content.

Processes

Processes are the building blocks of actions. They are defined as directional events occuring within and across different actor system fields, and containing mechanisms that operate upon input, throughput and state variables, producing values on throughput, state and output variables.

The processes used in the model to define the various actions are

- select action program, i.e. decide which action is to be performed next and at what time
- select message, i.e. compose evaluative and informational content
- offer message, involving selection of coactor to be activated
- expose to some activating coactor who is making a message offer
- attend to and interpret coactor's message, including noisy distortion of its content
- adapt internal actor states, such as interest, knowledge and evaluation
- choose, i.e. make a final commitment to either A or B.

Computer program structure

The above processes are declared as procedures in the computer program, and being called upon during the execution of actions making up interaction sequences. Although the procedures are called upon and executed sequentially in real world machine time, the system time is regulated by the dyadic interactive perspective, and held constant during a simulated interaction sequence. Thus the respective actions of two interactors may be regarded as occurring concurrently. The physical object blocks representing the various actors may switch between being in active and passive phases. This is made possible by the quasi-parallel processing abilities of the SIMULA language used.[1] In Fig. 1 this quasi-interactive execution of symbolic action programs by the various actor classes of the model is inadequately illustrated. The two upper boxes, representing two actor class person objects, show the left one to be carrying out a request of exchange program, and the right one to be complying.

Below is written the gross program description of these two actions.

line

```
I    requestexchange:  compose; send; if coactor == none
II                     then go to action program (1) else
III                    begin expose; attendinterpret; adapt; end;
IV   complyexchange:   expose; compose; send; if ¬accept
V                      then go to action program (1) else
VI                     begin attendinterpret; adapt; end;
```

A person being the current actor has selected a message using the compose procedure, after which the send procedure becomes activated (line I). This procedure involves his search for a suitable coactor among his group members. Let us assume that he has found one to forward a message to. Before the occurrence of any exposure to possible messages from the coactor, the latter takes over as the currently active. He has been activated by the actor and enters the complyingexchange sequence (line IV). Expose is the first procedure called upon. Before processing the received message, he will compose and send his own, which will be empty of content if the outcome of the expose procedure is such that he does not accept any message processing. Upon his forwarding the message—empty or not of content—the requesting actor is reactivated and goes through exposure, attention and interpretation of the message and adaptation to it, after which the coactor processes the original message if he has accepted it.

Through the above gross-level description of message exchange between two persons, we have illustrated an important feature of the model structure, made possible by the programming language used: Two instances of an actor class may co-exist and interact; and between two active phases belonging to one of the actors, active phases belonging to the coactor may occur.

Actor person states

The actual value at some point of time of state variables such as interest (evaluation strength above an interest threshold), accumulated knowledge and evaluative belief, choice and consistency, determine the

[1] Compare Birtwhistle et al. (1971).

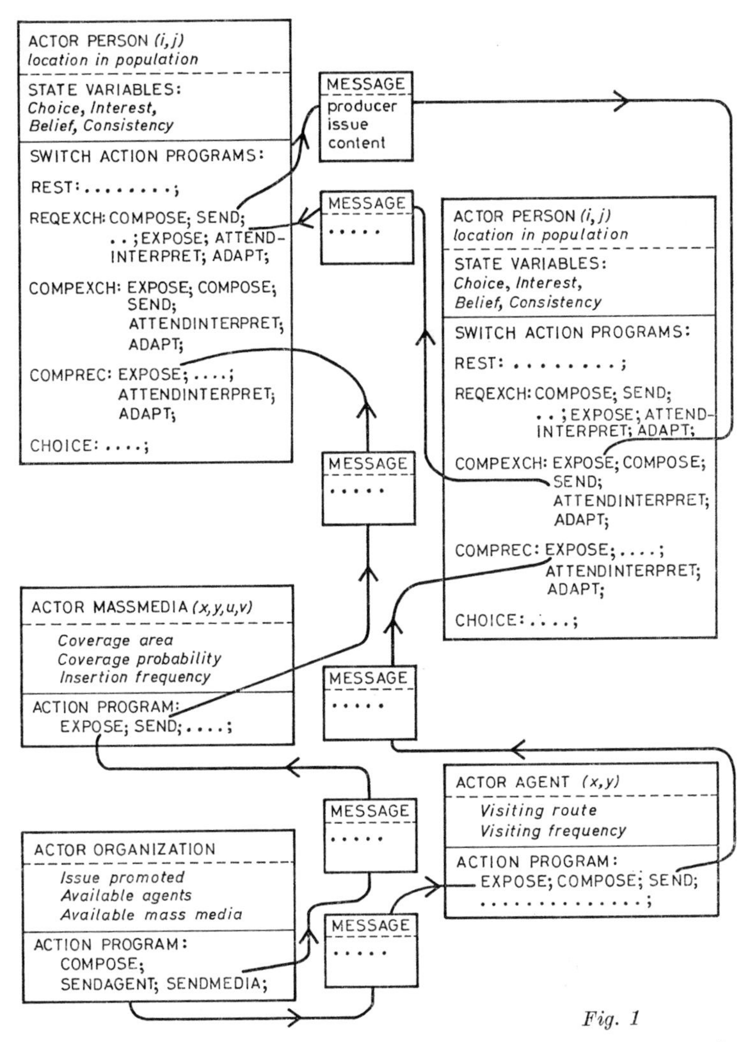

Fig. 1

person system state at that point of time. When states of cognitive consistency are included, the person may be treated as a ten-state system. For the present description and reduction modelling purpose, its treatment as a six-state system suffices.

The state of any one actor person $S(i, j)$ may then be denoted by the scalar variable $\eta_\tau(i, j)$ which at each instant of time τ takes on one of six integral values 1, 2, ..., 6, which are

$$\eta_\tau(i, j) = \begin{cases} 1: & \text{Neutral (not committed to any attitude or choice on the issue).} \\ 2: & \text{A-believer (with evaluative preference for A, without having committed any choice).} \\ 3: & \text{B-believer (with evaluative preference for B)} \\ 4: & \text{A-chooser (as an absorbing state).} \\ 5: & \text{B-chooser (as an absorbing state).} \\ 6: & \text{Fixed neutral (withdrawn to a state similar to 1, but one which is absorbing).} \end{cases} \tag{2.3}$$

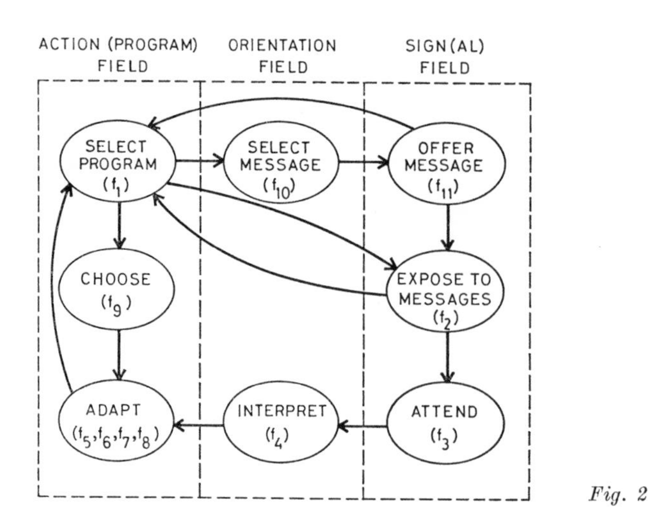

Fig. 2

The transitions allowed for between the above six states are given in Fig. 3 (in section 3).

Mechanisms and fields

Within the above structure mechanisms or operating characteristics may be specified, grouped according to the processes they belong to, and given computer-operational form through declaration of the procedures corresponding to these processes.

The mechanisms of the model represent a set of theoretical propositions, $f_1, f_2, \ldots, f_{11}$, concerning the adaptive, message processing and communicative behaviour of the actor person in relation to personal coactors, agents and mass media, and are interrelated according to the gross dyadic scheme illustrated in Fig. 2. This diagram represents processes and fields at the inter- and intra-personal level of the person actor in position as sender and in position as information-seeker or recipient.

Some of these mechanisms may be allocated to the *sign(al) field* (the right part of the diagram). They regulate the production and offering of sign compounds (f_{11}), overt information seeking and exposure behaviour (f_2), and attention (f_3) to sign compounds being exposed to.

Mechanisms regulating the selection of messages to be offered and the processing of messages produced by some coactors and paid attention to, are allocated to the *orientation field* (f_{10}, f_4). Mechanisms which have to do with the selection and execution of overt behaviour programs as well as covert adjustment programs are allocated to the *action (program) field* (the left part of the diagram). Internal states of cognition, evaluation, consistency and interest are regulated through adaptation (f_5, f_6, f_7, f_8). Overt behaviour programs concern non-symbolic actions and symbolic actions. Execution of the former means some act of choice between the issue alternatives (f_9). Selection of some program for symbolic actions (f_1) may lead either to production or seeking or reception of messages.

Actor resolution level

The notion of actor resolution level is crucial to interactor system simulation modelling; an object allows for gross and detailed descriptions in terms of number of actors, components, and relations without loosing the systems perspective of the whole, and simulations allow for protocol data that refer to levels equal to or grosser than the level at which the model mechanisms refer.

The set of mechanisms referred to above is formulated at the *intra-person* resolution level, while the simulation protocols used during the theory-explorative experiments with the model have been so constructed as to yield

- time series at the *aggregate* (population) level
- spatial patterns of distribution according to person actor states at the *interperson* (group) level
- state transition frequencies at the *person* level.

In the model system descriptions and reduction modelling to follow below the first- and last-listed levels will be referred to.

3. Model system description at person and aggregate levels

The person level

All the person actors $S(i, j)$ have similar structure and mechanisms. Environmental conditions for each actor during each time-period of unit length may be structured in discrete categories c, and we assume for the present descriptive purpose that the probability for each possible state-change between the six states from one time-point to the next time-point for constant factor values may be considered as dependent on c only. With also varying factor values taken into account, the possible transition probabilities may be summarized in

$$\mathbf{P}^c(\mathbf{x}) = (p^c_{rs}(\mathbf{x}))_{6\times 6}, \tag{3.1}$$

where

$$p^c_{rs}(\mathbf{x}) = \text{Prob}\ \{\eta_{\tau+1}(i, j) = s \,|\, \eta_\tau(i, j) = r;\ c,\ \mathbf{x}\} \tag{3.2}$$

and where $\mathbf{x}$ is a factor vector variable containing values of mechanism parameters.

The time-series of the state for an actor person may consequently be seen as a sequence of steps in a Markov-scheme; cf. Feller (1957). Such an application depends on the structuring of the conditions c into proper categories. Of the categories used for possible situations for an actor $S(i, j)$ during time-periods between two consequtive integral time-points the following are identified here:

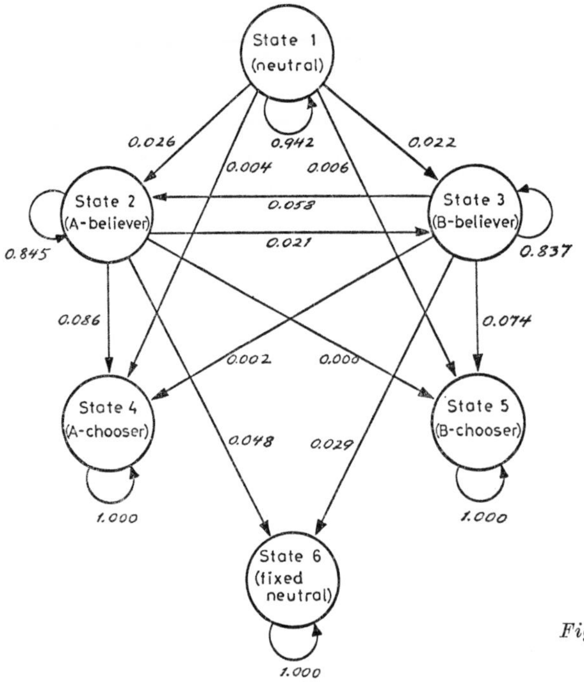

Fig. 3

Condition		c
undifferentiated		0
the state of the "membership group" (2.2)	none of group members has state 4 or 5 . .	1
	at least one group member has state 4 but none has state 5	2
	at least one group member has state 5 but none has state 4	3
	at least one group member has state 4 and at least one has state 5	4

(3.3)

Estimates of transition probabilities are obtained by calculating the corresponding transition frequencies. Thus, from each simulation run estimates of the matrices (3.1) for the conditions c may be obtained. As an illustration, the mean of the ten estimates on the matrix $\mathbf{P}^{\circ}(\mathbf{0})$ from the ten simulations in an experiment series (3.8) is given in the above diagram (Fig. 3), which also shows the transitions allowed for in the person as a six-state system.

The above estimate of the transition matrix illustrates some general characteristics of state-changes being true for all conditions c and for all factor values $\mathbf{x}$.

The aggregate level

The connection of the 100 states $\eta_\tau(i, j)$ at the person resolution level with the state of the system described at the aggregate resolution level may be given by the identity

$$\boldsymbol{\eta}_\tau = 0.01 \sum_{i,j=1}^{10} \boldsymbol{\iota}_\tau(i, j), \tag{3.4}$$

where

$$\iota_{r\tau}(i, j) = \begin{cases} 1 & \text{if} \quad \eta_\tau(i, j) = r \\ 0 & \text{otherwise,} \end{cases} \tag{3.5}$$

i.e. at the aggregate level the state of the system is the vector of proportions of person actors being in the six possible person states.

The behaviour of the aggregate population may be defined in terms of the state of the system at time-points τ

$$\boldsymbol{\eta}_\tau = \begin{pmatrix} \eta_{1\tau} \\ \vdots \\ \eta_{6\tau} \end{pmatrix} \tag{3.6}$$

The system behaviour may be described as to allow the following representation

$$\begin{cases} \boldsymbol{\eta}_{\tau+1} = \boldsymbol{\varkappa}(\mathrm{x}, \boldsymbol{\eta}_\tau) + \boldsymbol{\delta}_{\tau+1}(\mathrm{x}, \boldsymbol{\eta}_\tau) \\ E(\boldsymbol{\eta}_{\tau+1} \,|\, \mathrm{x}, \boldsymbol{\eta}_\tau) = \boldsymbol{\varkappa}(\mathrm{x}, \boldsymbol{\eta}_\tau), \end{cases} \tag{3.7}$$

where $\boldsymbol{\varkappa}$ is a vector function, x is the factor vector variable and $\boldsymbol{\delta}_{\tau+1}$ is a vector of disturbances. For purposes of output analysis, the promotional inputs to the population are regarded as a fixed set which is evenly spread out over time and not subject to variation. Thus it is omitted in the above relation system.

The kind of model system behaviour being generated at this level is exemplified in Figs. 4 and 5. Fig. 4 shows ten time-series of the development of proportion of neutrals, $y_{1\tau}$, in a series of simulations, in which the system has been released from the initial aggregate state

$$\mathrm{y}_0^I = \begin{pmatrix} 0.95 \\ 0.00 \\ 0.00 \\ 0.05 \\ 0.00 \\ 0.00 \end{pmatrix} \tag{3.8}$$

(with this initial state as the marginal distribution, when the system is studied at the person resolution level; cf. (3.4–5)) for constant factor values held at the centre point ($\mathbf{x} = 0$).

Fig. 5 gives some time series of the development of proportions Neutrals, A-choosers, and B-choosers from one run belonging to another series of ten simulations from the initial state

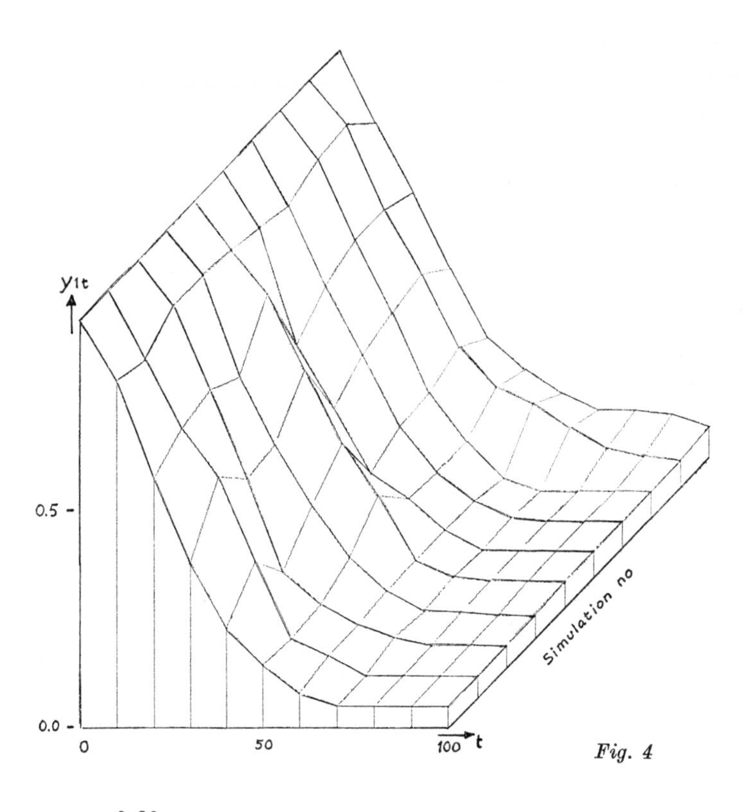

Fig. 4

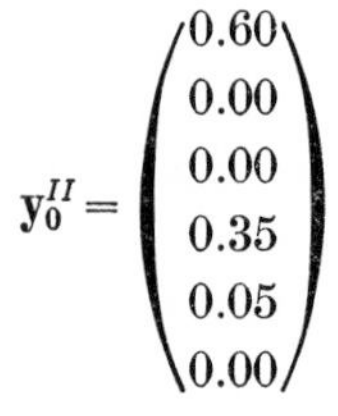

$$\mathbf{y}_0^{II} = \begin{pmatrix} 0.60 \\ 0.00 \\ 0.00 \\ 0.35 \\ 0.05 \\ 0.00 \end{pmatrix} \tag{3.9}$$

for constant factor values held at the centre point.

In the above two series variation has been made only in the values of

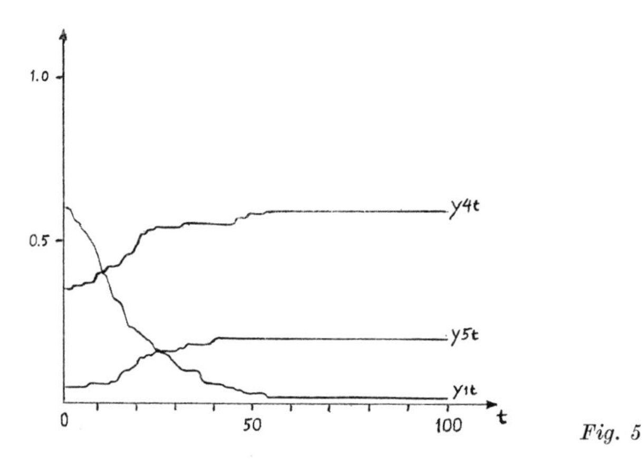

Fig. 5

so called "stream parameters", generating the streams of pseudo-random numbers used in the random drawings employed in the mechanisms. While the initial chooser proportions are different for the two series, promotional aggregate inputs are the same: Mass media are used by the organization promoting the A-issue, while the competing organization employs agent promotion. Protocol data from these two series will be used in the below reduction modelling of the model.

4. Reduced models at the aggregate level[1]

A characteristic shared by the above model with many models of human interaction, formulated at a micro-level, is the high degree of complexity. It is this feature that necessitates simulations in order that its dynamic implications may be studied—and in order that feedback information be obtained through procedures such as sensitivity analyses[2], which can be used in attempts at regulating or reducing the model in terms of (i) mechanisms, (ii) variables and relations, and (iii) the actor resolution level referred to. If a model alternative is built at a grosser level of resolution on the basis of simulation output of the original, this becomes a reduced model of the latter. It is this kind of reduction modelling that will be turned to in the following. Two models will be given with reference to the aggregate level, followed by a third model formulated at the person level.

The problem of substituting the original complex model with a system of relations between variables, constituting a reduced model, is simplified inasmuch as the factors (mechanism parameters) are assumed to be constant with values lying in the centre point.

Model number 1

In the first reduction experiment it is assumed that the state vector (3.6) follows an autoregressive scheme of the first order

---

[1] This and the following section are based on Bråten & Norlén (1969), Norlén (1970) and Bråten & Norlén (1972*a*).

[2] Partly for economic reasons, this kind of analysis is not without problems, cf. Bråten & Norlén (1972*b*). In a sensitivity analysis on a subset of the model mechanism parameters, 12 factors were selected, using a fractional factorial design of 128 experimental points. Sensitivity estimates were obtained from a population of (i) state transition probabilities (3.2) and aggregate system variables (3.6). The regression equations employed are polynomials of the second degree in the factors,

$$\mu + \boldsymbol{\beta}'\mathbf{x} + \mathbf{x}'\boldsymbol{\Gamma}\mathbf{x} \tag{4.1}$$

which provide for estimation of mean effects ($\boldsymbol{\beta}$) and interaction effects of the second order ($\boldsymbol{\Gamma}$). The analysis provides a rather poor feedback in terms of model substance, but is a rich source for illustrations of problems of methods.

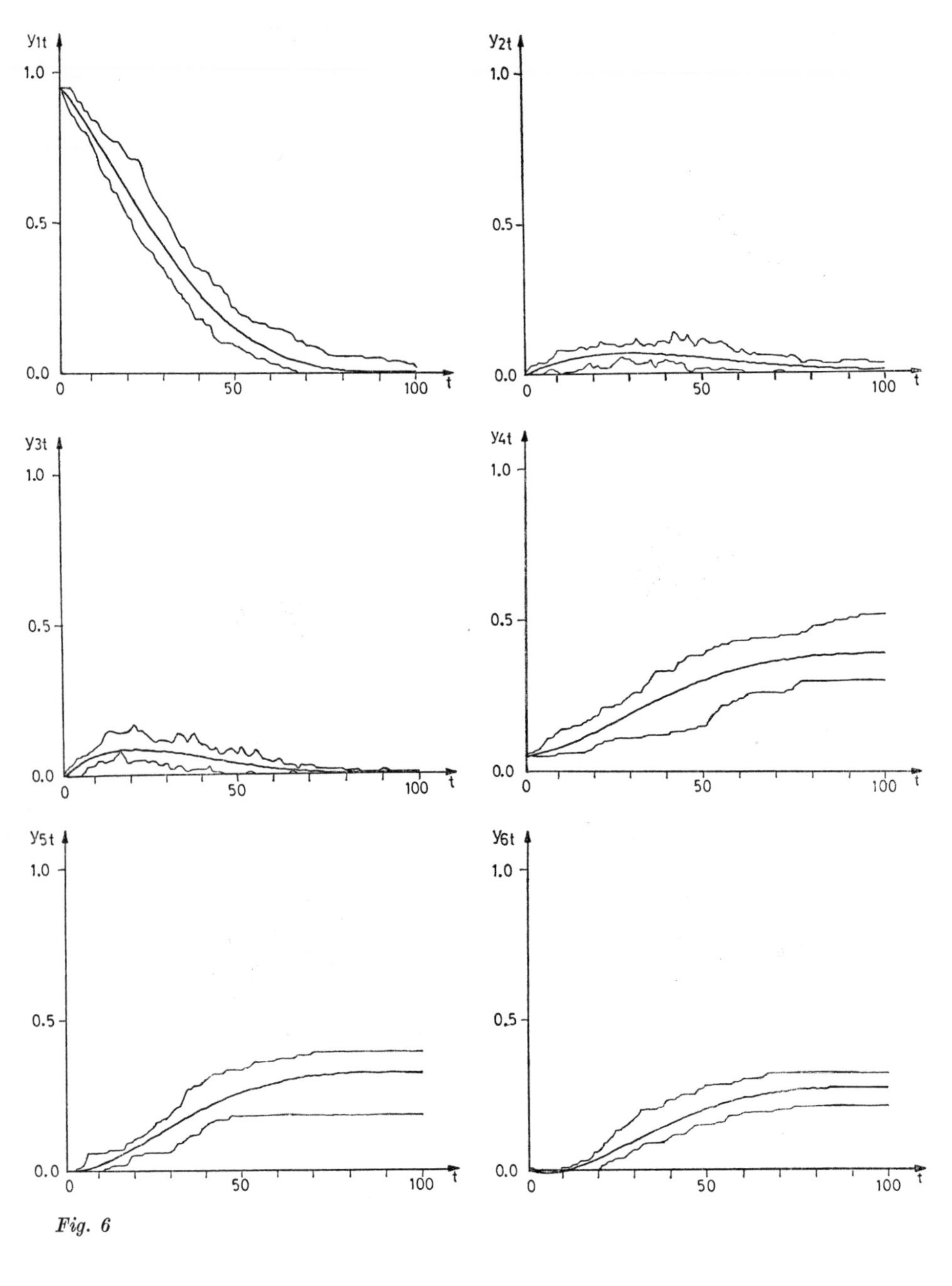

Fig. 6

$$\begin{cases} \boldsymbol{\eta}_{\tau+1} = \mathbf{B}\boldsymbol{\eta}_\tau + \boldsymbol{\delta}_{\tau+1} \\ E(\boldsymbol{\eta}_{\tau+1} \mid \boldsymbol{\eta}_\tau) = \mathbf{B}\boldsymbol{\eta}_\tau. \end{cases} \tag{4.2}$$

cf. (3.7).

A suitable method for the evaluation of the reduction experiment is on the one hand to use material from the first-listed experiment series

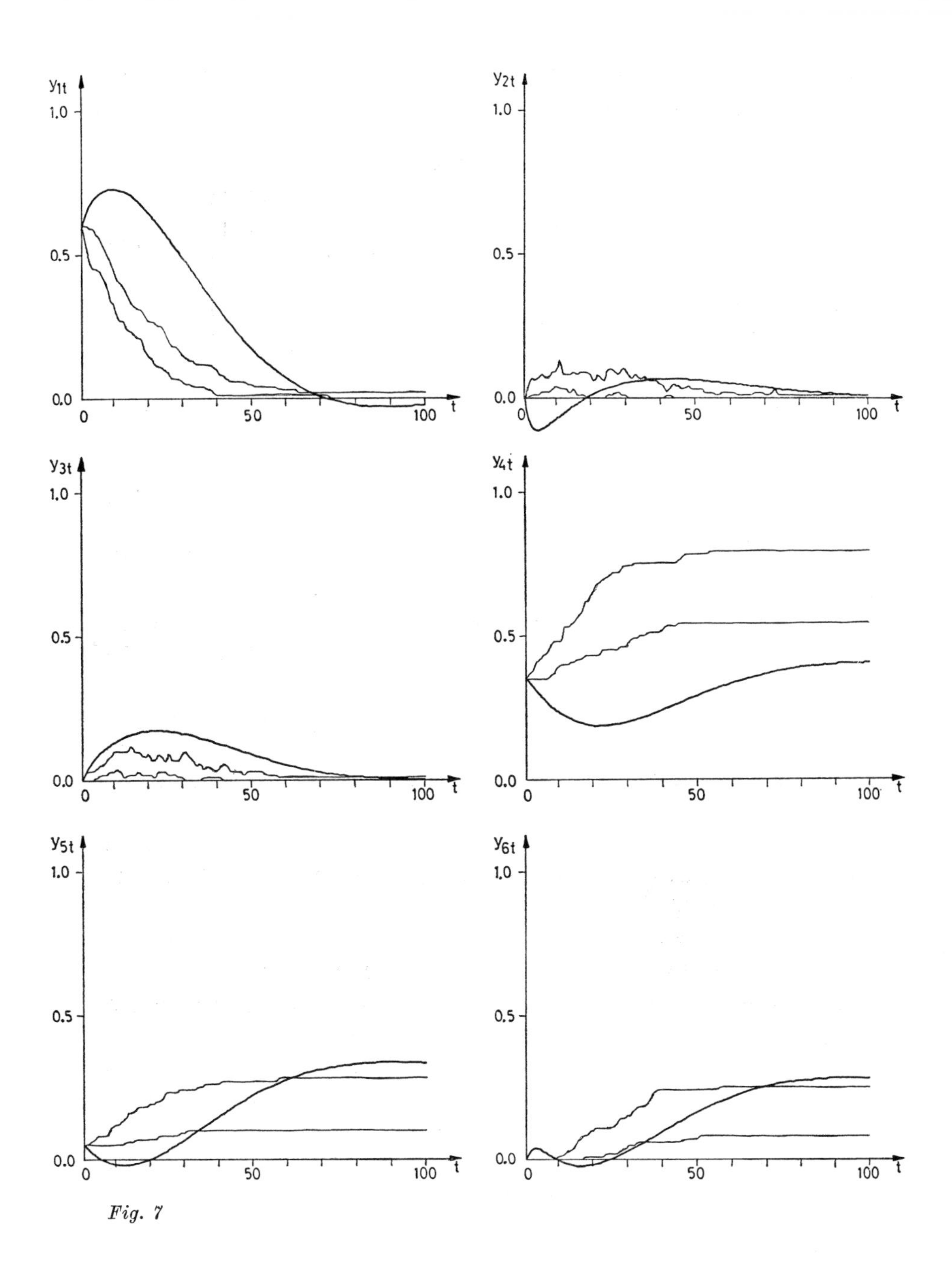

Fig. 7

released from the state $\mathbf{y}_0^I$, for the estimation of the reduced model and, on the other hand, to use the other series released from the state $\mathbf{y}_0^{II}$ for investigation of the suitability of the approach.

The coefficient matrix B is estimated from the calculated time-series of the mean values for the state variables from the ten time-series from the former experiment series. Application of the method of least squares, relation for relation, gives the following estimate[1]

$$\mathsf{B}=\begin{bmatrix} 0.976 & 0.024 & -0.190 & 0.171 & -0.095 & -0.130 \\ 0.019 & 0.690 & 0.267 & -0.249 & 0.152 & 0.183 \\ 0.008 & -0.019 & 0.895 & 0.062 & 0.056 & -0.153 \\ 0.004 & 0.057 & 0.036 & 0.951 & 0.007 & 0.061 \\ 0.000 & 0.095 & 0.076 & -0.028 & 0.855 & 0.206 \\ -0.007 & 0.153 & -0.083 & 0.093 & 0.025 & 0.833 \end{bmatrix} \tag{4.3}$$

The system (4.2) gives the following predictions for the mean value of the state variables

$$\bar{\mathbf{y}}_t^p = \mathsf{B}^t \mathbf{y}_0. \tag{4.4}$$

Fig. 6 shows predicted time-series from the initial value $\mathbf{y}_0^I$ together with the ranges of variation as were obtained from the simulations.

As may be seen, the picture shows good agreement in this case, as was to be expected, since the material for comparison is taken from the same material as formed the basis for estimation.

However, the picture is quite another when predictions are made for the other experiment series. Fig. 7 shows predictions (4.4) from the initial value $\mathbf{y}_0^{II}$ together with the ranges of variation from this experiment series.

The Figure shows big differences as compared with the simulation results obtained. The conclusion is that the relation-system (4.2) is an all too crude simplification.

Although the experiment proved negative, the study is nevertheless interesting inasmuch as the relation system (4.2) is an example of a frequently used formulation of dynamic processes in econometrics which is characterized by, inter alia, the assumption of a time-invariant system-structure; see e.g. Johnston (1963). The method applied thus indicates a course which permits of investigation of the previously deduced usual ways of specifying relation models. With the aim of further elucidating this argument the problem of replacement and model formulation will be illustrated with another experiment.

Model number 2

Also with this experiment the system (4.2) is used as point of departure. The following supplementary assumptions are added:

---

[1] It may be noticed that the column sums of (4.3) are all equal to 1. This property is generally valid for this kind of regression, see Lee et al. (1970).

$$\begin{cases} 0 \leqslant \beta_{ij} \leqslant 1 & i, j = 1, \ldots, 6 \\ \sum_{i=1}^{6} \beta_{ij} = 1 & j = 1, \ldots, 6 \\ \beta_{61} = \beta_{12} = \beta_{13} = 0 \\ \beta_{44} = \beta_{55} = \beta_{66} = 1 \end{cases} \tag{4.5}$$

These restrictions imposed on the parameters are obtained from a priori knowledge of the nature of the system at the person resolution level, in which connection the coefficient matrix is assumed to follow the same restrictions as apply for the transposed matrix (3.1) with transitional probabilities, i.e.

$$\mathsf{B} \sim \mathbf{P}(\mathbf{0})'. \tag{4.6}$$

From the econometric point of view also this case is of interest, since in econometrics restrictions are often imposed on parameter values in the relation structures.

The use of the principle of least squares transfers the estimation question to a quadratic programming problem; cf. Lee et al. (1970). The following estimate is obtained:

$$\mathsf{B} = \begin{bmatrix} 0.973 & — & — & — & — & — \\ 0.007 & 0.891 & 0.045 & — & — & — \\ 0.014 & 0.000 & 0.890 & — & — & — \\ 0.004 & 0.048 & 0.000 & 1.000 & — & — \\ 0.002 & 0.021 & 0.042 & — & 1.000 & — \\ — & 0.040 & 0.023 & — & — & 1.000 \end{bmatrix}, \tag{4.7}$$

where the parameter values assumed to be equal to zero are marked with a dash.

Fig. 8 shows predicted time-series (4.4) from the initial state $\mathbf{y}_0^I$ with the use of $\mathbf{B}$ from (4.7).

If the time-series from this experiment are compared with the time-series from the autoregressive system without restrictions, reproduced in Fig. 6, Fig. 8 shows a poorer agreement in relation to the simulation results obtained. This is explained in the first place by the fact that the system with restrictions has fewer degrees of freedom, which gives a poorer "goodness-of-fit".

If the predictions are made for the second experiment series (see Fig. 9), however, one gets a comparatively better result than that obtained from the first experiment; cf. Fig. 7.

If we sum up the results in this section we get the general principles that one should endeavour if possible to fit a priori knowledge of the system into formal abstractions, and that goodness-of-fit criteria based on the same material as that used for estimation, do not give assurances of the suitability of the approach in other situations.

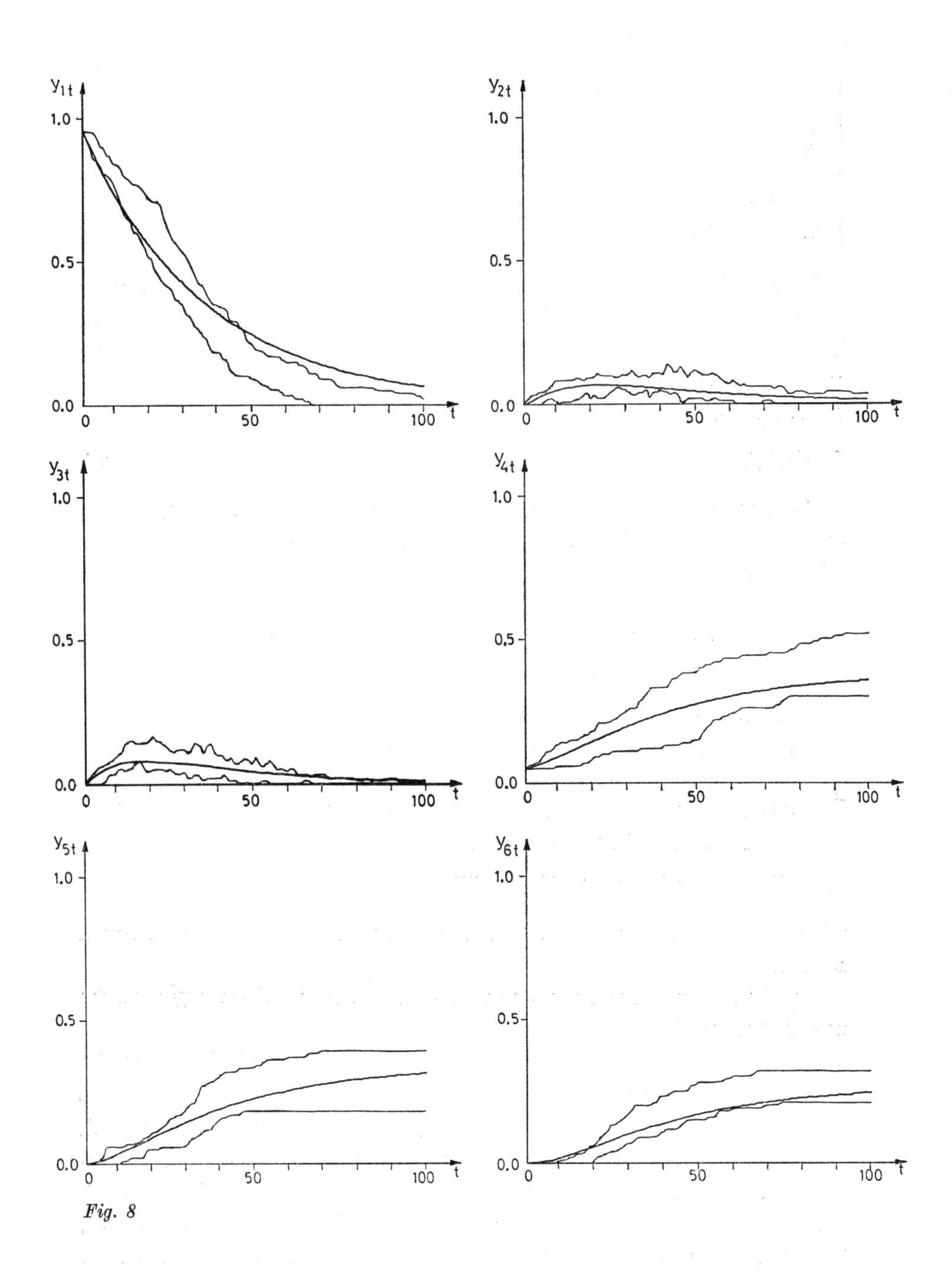

Fig. 8

5. Reduced models at the person level

In this section the task is to try, on the basis of knowledge of the system at the person resolution level, to build up a set of relations that give the behaviour of the system at the aggregate resolution level.

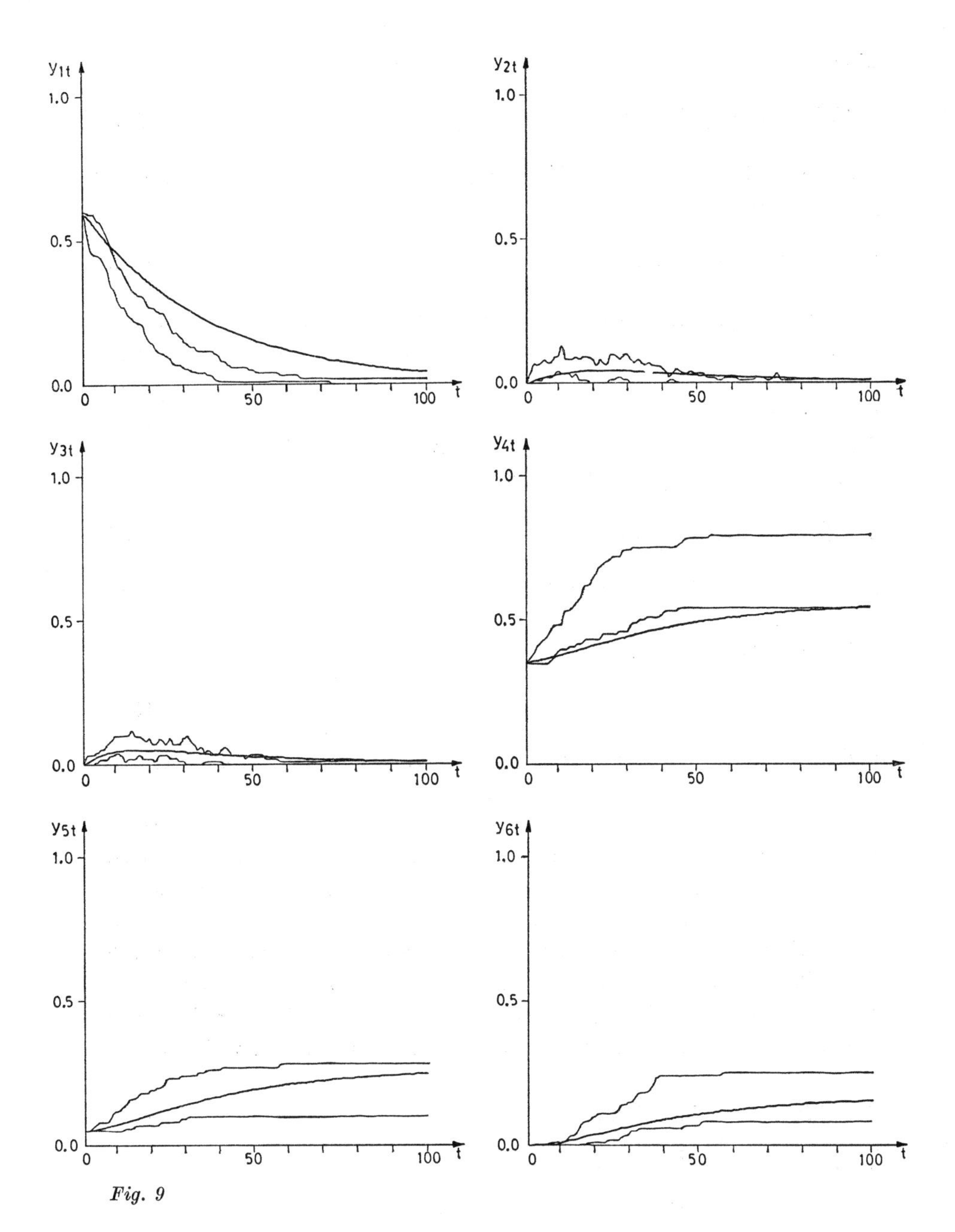

Fig. 9

Model number 3

Considered stochastically, the state variable for the person actor $\mathbf{S}(i, j)$ may be written

$$\mathbf{p}_\tau(i, j) = \begin{pmatrix} p_{1\tau}(i, j) \\ \vdots \\ p_{6\tau}(i, j) \end{pmatrix}, \tag{5.1}$$

where

$$p_{r\tau}(i, j) = \text{Prob}\,\{\eta_\tau(i, j) = r\}. \tag{5.2}$$

From section 3 it emerges that the initial states are distributed with a given marginal distribution, i.e.

$$\mathbf{p}_0(i, j) = \mathbf{y}_0 \quad i, j = 1, \ldots, 10. \tag{5.3}$$

Let us assume that the effect that depends on inputs from the environment common to all person actors may be neglected. Let us further assume that the effect of the circumstance that certain actors have fewer potential coactors (border cases) may also be neglected. From this it follows that, since all person actors have the same structures it will hold true for every point of time τ that

$$\mathbf{p}_\tau(i, j) = \mathbf{p}_\tau \quad i, j = 1, \ldots, 10, \tag{5.4}$$

i.e. the probabilities that an actor is in some one of the six states are the same for all actors.

From (5.3–4) it follows that

$$E(\boldsymbol{\eta}_\tau) = \mathbf{p}_\tau \tag{5.5}$$

and $\mathbf{p}_\tau$ may thus be used as an unbiassed estimator for the state of the system at the person resolution level.

One way of solving the reduction problem would therefore be to try to get a system of relations that binds together probabilities (5.5) for consecutive points of time. As an aid here we have the estimates of the transition matrices referred to in section 3.

In this connection it is assumed that the transition matrices for the mutually exclusive conditions $c = 1, 2, 3, 4$ are relevant. Let us consider a person actor and his neighbours (2.2) at the point of time τ. The four probabilities that the conditions c will occur are calculated with

Condition c	*Probability*
1	$w^1(\mathbf{p}_\tau) = \pi_{400}$
2	$w^2(\mathbf{p}_\tau) = \pi_{040} + \pi_{130} + \pi_{220} + \pi_{310}$
3	$w^3(\mathbf{p}_\tau) = \pi_{004} + \pi_{103} + \pi_{202} + \pi_{301}$
4	$w^4(\mathbf{p}_\tau) = \pi_{013} + \pi_{022} + \pi_{031} + \pi_{112} + \pi_{121} + \pi_{211}$

(5.6)

where

$$\pi_{ijk} = \binom{4}{i}\binom{4-i}{j}(1 - p_{4\tau} - p_{5\tau})^i p_{4\tau}^j p_{5\tau}^k. \tag{5.7}$$

The probability that the actor system will be in state s at the point of time τ will thus be

$$p_{s\tau+1} = \sum_{c=1}^{4} w^c(\mathbf{p}_\tau) \sum_{r=1}^{6} p_{r\tau} p^c_{rs}(0), \tag{5.8}$$

where $\mathbf{p}^c_{rs}(\mathbf{o})$ is the probability for transition from state r to state s.

The whole probability vector may be written in the following compact way

$$\mathbf{p}_{\tau+1} = \left[\sum_{c=1}^{4} w^c(\mathbf{p}_\tau)\mathbf{P}^c(\mathbf{0})\right]' \mathbf{p}_\tau \tag{5.9}$$

But (5.9) applies for all actor systems. It therefore follows that

$$E(\boldsymbol{\eta}_{\tau+1}) = \left[\sum_{c=1}^{4} w^c(E(\boldsymbol{\eta}_\tau))\mathbf{P}^c(\mathbf{0})\right]' E(\boldsymbol{\eta}_\tau). \tag{5.10}$$

Thus with estimates of the $\mathbf{P}^c(\mathbf{0})$:s one can recursively get predictions from (5.10) of the expectation of the state for the system at the aggregate resolution level.

As an estimate of the conditioned transition matrices $\mathbf{P}^c(\mathbf{0})$ we use the calculated frequencies from experimental series released from the state $\mathbf{y}^I_0$.

In the same way as in the previous section we calculate predictions from the two initial states $\mathbf{y}^I_0$ and $\mathbf{y}^{II}_0$. In the two Figs. 10 and 11 are shown the time-series obtained together with variation ranges from the simulation results actually obtained from the two experiment series referred to in section 3.

As emerges from the figures, the predictions do not show structural resemblances throughout to simulation results obtained.[1] These differences are probably ascribable to the all too simplified assumptions; this is illustrated by, inter alia, Fig. 10, which shows that not even the predictions based on the same material as that serving as a basis for estimation show agreement.

---

[1] If the reduced models of the form (4.2) or (5.9) had been successful, one might have introduced tests for variable factor values. For the former system could be used

$$\begin{cases} \boldsymbol{\eta}_{\tau+1} = \mathsf{B}(\mathbf{x})\boldsymbol{\eta}_\tau + \boldsymbol{\delta}_{\tau+1} \\ E(\boldsymbol{\eta}_{\tau+1} \mid \mathbf{x}, \boldsymbol{\eta}_\tau) = \mathsf{B}(\mathbf{x})\boldsymbol{\eta}_\tau \end{cases} \tag{5.11}$$

with $\mathbf{B}(\mathbf{x})$ approximated with e.g. an expression in the first or second degree is $\mathbf{x}$.

For system (5.9) it seems that variation in the factor value $\mathbf{x}$ may be introduced through

$$\mathbf{p}_{\tau+1} = \left[\sum_{c} w^c(\mathbf{p}_\tau)\mathbf{P}^c(\mathbf{x})\right]' \mathbf{p}_\tau \tag{5.12}$$

in which connection approximations of $\mathbf{P}^c(\mathbf{x})$ are resorted to; cf. foot-note 2 on p. 159.

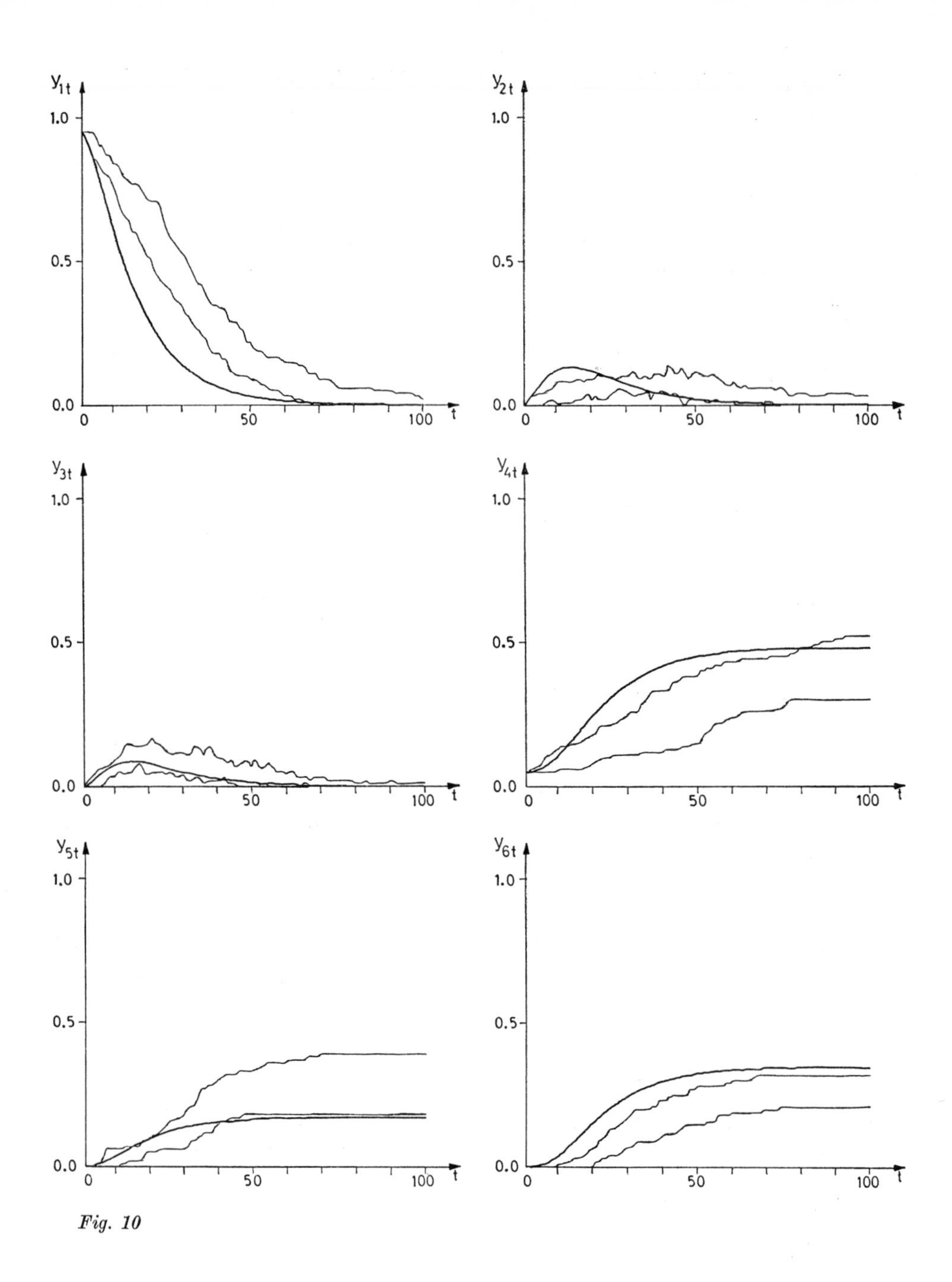

Fig. 10

The underlying train of thought in the reduction experiment does, however, indicate a more or less generally applicable principle. With the aim of presenting the main line in the train of thought the following example may be given, where the system (5.9) in a special case is shown to contain the simple logistic system.

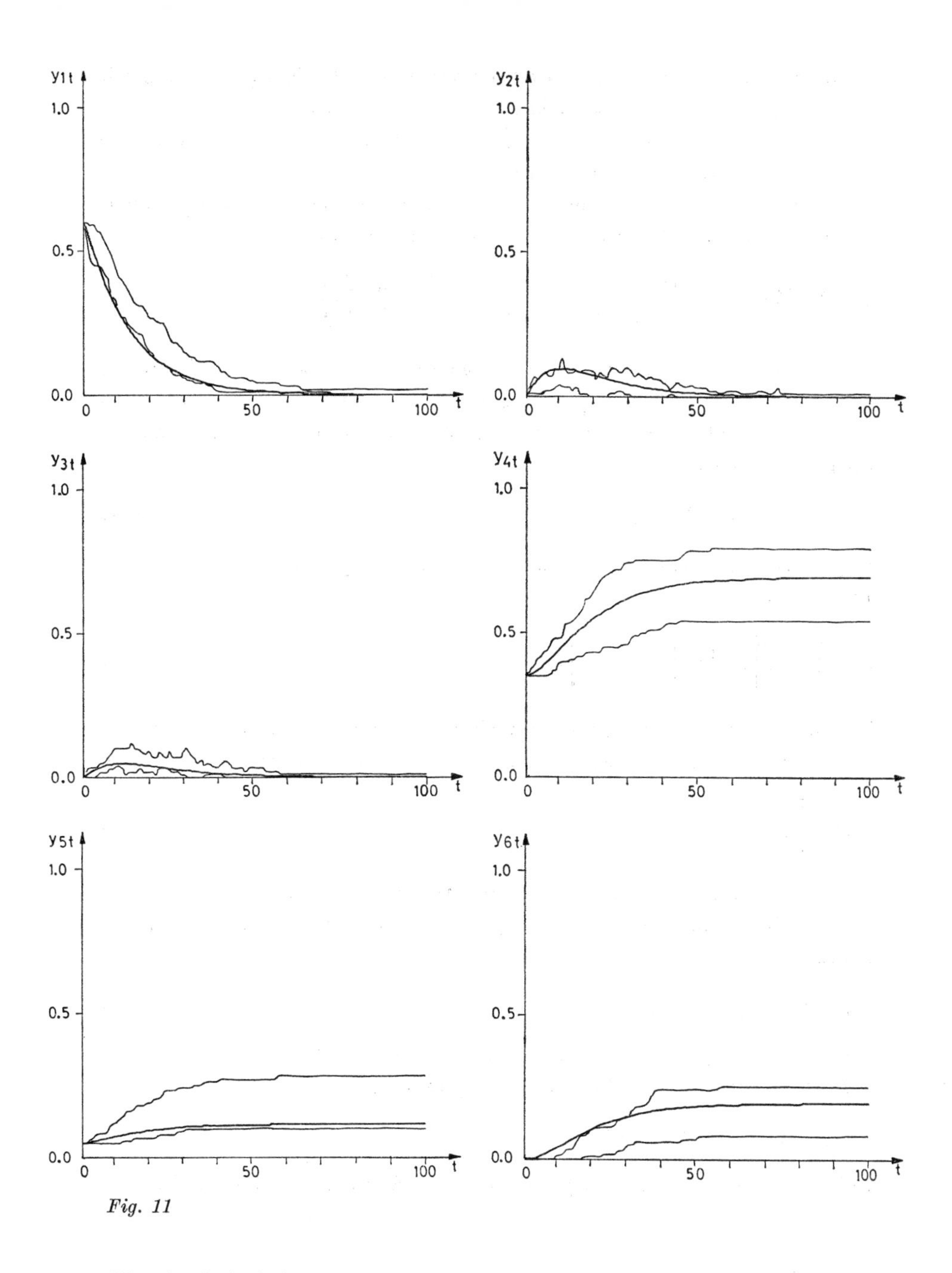

Fig. 11

The simple logistic system

Let us consider the logistic system, as e.g. described by Karlsson (1958). Such a system may be viewed as consisting of a large number of actors which all have the same structure. At every point of time each actor system assumes either the state-value 0 or the state-value 1. Each of them interacts once during every period of time from τ to $\tau+1$ with a randomly chosen coactor. If this coactor has state 0, the actor state is not changed. If, on the other hand, the coactor has state 1 and the actor is in state 0, the state of the actor system is changed to 1 with the probability α. State 1 is an absorbing state.

From the description it follows that the probabilities that a person is in one of the two states are the same for all persons, i.e.

$$\mathbf{p}_\tau = \begin{pmatrix} p_{1\tau} \\ p_{2\tau} \end{pmatrix} = \begin{matrix} \text{the probability that at subsystem at} \\ \text{point of time } \tau \text{ is in state} \end{matrix} \begin{pmatrix} 0 \\ 1 \end{pmatrix} \tag{5.13}$$

if the probability vector $\mathbf{p}_0$ for the initial states is the same for all persons.

The probabilities that a person actor will interact with a coactor with state 0 or state 1 during a period of time from τ to $\tau+1$ can be approximated with

$$w^1 = p_{1\tau} \quad \text{and} \quad w^2 = p_{2\tau} \tag{5.14}$$

respectively. The two matrices with the transition-probabilities can further be written

$$\mathbf{P}^1 = \begin{bmatrix} 1 & 0 \\ 0 & 1 \end{bmatrix}, \quad \mathbf{P}^2 = \begin{bmatrix} 1-\alpha & \alpha \\ 0 & 1 \end{bmatrix}. \tag{5.15}$$

The weighting of these matrices with their probabilities of occurring as weights gives the following system of relations for the probability vector

$$\mathbf{p}_{\tau+1} = \left[\sum_{c=1}^{2} w^c(\mathbf{p}_\tau) \mathbf{P}^c \right]' \mathbf{p}_\tau. \tag{5.16}$$

The verbal description agrees with the assumptions for the logistic system. In a discrete form the logistic system may be written in the following way

$$\eta_{\tau+1} = \eta_\tau + \alpha\eta_\tau(1-\eta_\tau), \tag{5.17}$$

where η_τ corresponds to the probability $p_{2\tau}$ in this example.

If the second row in (5.16) is developed we get

$$p_{2\tau+1} = p_{2\tau}(\alpha p_{1\tau} + p_{1\tau} + p_{2\tau}), \tag{5.18}$$

which with the substitution $p_{1\tau} = 1 - p_{2\tau}$ is reduced to the form (5.17).

6. Summary

In this paper the structure of a model, used for theory-explorative simulations in the intersection of human communication theory and theories of cognitive consistency, has been described and subjected to modelling attempts, aimed at model-descriptive vehicles formulated at grosser levels of resolution than that of the original.

In the description, some of the basic concepts of an interaction system modelling framework[1] have been illustrated, and the usage of the programming language applied has been indicated. While tenability evaluation analyses and theory-explorative outputs of the model have been given elsewhere (Bråten 1970, 1972), sections 4 and 5 of the present paper have been devoted to attempts at building reduced models of the model. Two such models with reference to the aggregate level, turned out to be too gross descriptive tools in terms of reproduction of aggregate behaviour. A third attempt, utilizing Markovian transition probabilities generated by simulations with the original, showed better results. While not acceptable, they can at least be used to indicate that this manner of *complex model-reduced model interactive simulations* may be of help in search of elegancy and simplicity when faced with a complex system.

---

[1] The interaction modelling framework has also been utilized in the SIMTEST model (Norlén 1972) and in the Dyadsimulator model (Bråten 1971, 1972).

References to the Appendix

Abelson, R. P. & Bernstein, A. (1963). A computer simulation model of community referendum controversies. *Public Opinion Quarterly, 27*, 93–122.

Birtwhistle, G., Dahl, O. J., Myhrhaug, B. & Nygaard, K. (1971). *SIMULA begin*. Lund: Studentlitteratur (in press).

Bråten, S. (1968*a*). Marknadskommunikation. Stockholm: Beckmans.

—— (1968*b*). A simulation study of personal and mass communication. *IAG Quaterly 2*, 7–28. Amsterdam: IFIP Administrative Data Processing Group. Also in Stockhaus (ed.) (1970): *Models and simulation*. Gothenburg: Akademiförlaget/Scand. Univ. Books.

—— (1968*c*) *Progress report on the SIMCOM model*. Solna: Institute of Market and Societal Communication.

—— (1970). Communication mechanisms. *IMAS Information*, Report no. 26E1170, Solna.

—— (1971). The human dyad. *IMAS Information*, Report no. 24E0271, Solna.

—— (1972). Theory and simulation of communication and consistency. Inst. of Psychology, Univ. of Bergen (in preparation).

Bråten, S. & Norlén, U. (1969). Simulation model analysis. *IMAS Information*, Report no. 11E0269, Solna.

—— (1972*a*). Simulation model analysis and reduction. In Computer Simulation Versus Analytical Solutions for Business and Economic Models. Symposium working papers, vol. I, Univ. of Gothenburg.

—— (1972*b*). Sensitivity analysis of computer simulation models (in preparation).

Dahl, O. J. & Nygaard, K. (1965). *SIMULA—a language for programming and description of discrete event systems. Introduction and user's manual*. Oslo: Norwegian Computing Center.

Feller, W. (1957). *Introduction to probability theory and its applications*. Vol. I. New York: Wiley.

Gullahorn, J. T. & Gullahorn, J. E. (1972). Social and cultural system simulation. In H. Guetzkow, P. Kotler & R. Schultz (eds.): *Simulation in Social and Administrative Science*. Englewood Cliffs: Prentice-Hall.

Hägerstrand, T. (1953). *Innovationsförloppet ur korologisk synpunkt*. Lund: Glerup.

Johnston, J. (1963). *Econometric methods*. 2nd ed. 1972. New York: McGraw-Hill.

Karlsson, G. (1958). *Social mechanisms*. Stockholm: Almqvist & Wiksell.

Lee, T. C., Judge, G. G. & Zellner, A. (1970). *Estimating the parameters of the Markov probability model from aggregate time series of data*. Amsterdam: North-Holland.

Norlén, U. (1970). Om specifikation av relationsmodeller. Dept. of Statistics, Univ. of Uppsala.

—— (1972). *Simulation model building*. Stockholm: Almqvist & Wiksell.

Starbuck, W. H. & Dutton, J. M. (1972). Computer simulation as a tool for descriptive behavioral science. In Computer Simulation versus Analytical Solutions for Business and Economic Models, Symposium Working papers, vol. I, Univ. of Gothenburg.